Industry 4.0

Manufacturing Design and Technology Series

Series Editor: J. Paulo Davim, University of Aveiro, Portugal

This series will publish high quality references and advanced textbooks in the broad area of manufacturing design and technology, with a special focus on sustainability in manufacturing. Books in the series should find a balance between academic research and industrial application. This series targets academics and practicing engineers working on topics in materials science, mechanical engineering, industrial engineering, systems engineering, and environmental engineering as related to manufacturing systems, as well as professions in manufacturing design.

Drills
Science and Technology of Advanced Operations
Viktor P. Astakhov

Technological Challenges and Management
Matching Human and Business Needs
Edited by Carolina Machado and J. Paulo Davim

Advanced Machining Processes
Innovative Modeling Techniques
Edited by Angelos P. Markopoulos and J. Paulo Davim

Management and Technological Challenges in the Digital Age
Edited by Pedro Novo Melo and Carolina Machado

Machining of Light Alloys
Aluminum, Titanium, and Magnesium
Edited by Diego Carou and J. Paulo Davim

Additive Manufacturing
Applications and Innovations
Edited by Rupinder Singh and J. Paulo Davim

Emotional Intelligence and Neuro-Linguistic Programming
New Insights for Managers and Engineers
Edited by Carolina Machado and J. Paulo Davim

Business Intelligence and Analytics in Small and Medium Enterprises
Edited by Pedro Novo Melo and Carolina Machado

Enabling Technologies for the Successful Deployment of Industry 4.0
Edited by Antonio Sartal, Diego Carou and J. Paulo Davim

Industry 4.0: Challenges, Trends, and Solutions in Management and Engineering
Edited by Carolina Machado and J. Paulo Davim

For more information about this series, please visit: https://www.crcpress.com/Manufacturing-Design-and-Technology/book-series/CRCMANDESTEC

Industry 4.0

Challenges, Trends, and Solutions in Management and Engineering

Edited by Carolina Machado and J. Paulo Davim

CRC Press
Taylor & Francis Group
Boca Raton London New York

CRC Press is an imprint of the
Taylor & Francis Group, an **informa** business

CRC Press
Taylor & Francis Group
6000 Broken Sound Parkway NW, Suite 300
Boca Raton, FL 33487-2742

Printed on acid-free paper

International Standard Book Number-13: 978-0-8153-5440-6 (Hardback)
International Standard Book Number-13: 978-0-367-49483-4 (Paperback)

Library of Congress Cataloging-in-Publication Data

Names: Machado, Carolina, 1965- editor. | Davim, J. Paulo, editor.
Title: Industry 4.0 : challenges, trends, and solutions in management and engineering / edited by Carolina Machado and J. Paulo Davim.
Other titles: Industry 4.0 (CRC Press) | Industry four point zero
Description: Boca Raton, FL : CRC Press/Taylor & Francis Group, [2020] | Series: Manufacturing design and technology | Includes bibliographical references and index.
Identifiers: LCCN 2020003335 (print) | LCCN 2020003336 (ebook) | ISBN 9780815354406 (hardback ; acid-free paper) | ISBN 9781351132992 (ebook) | ISBN 9781351132985 (adobe pdf) | ISBN 9781351132961 (mobi) | ISBN 9781351132978 (epub)
Subjects: LCSH: Computer integrated manufacturing systems. | Production management--Data processing. | Manufacturing industries--Technological innovations. | Internet of things. | Business enterprises--Computer networks. | Automation.
Classification: LCC TS155.63 .I369 2020 (print) | LCC TS155.63 (ebook) | DDC 658.5/14--dc23
LC record available at https://lccn.loc.gov/2020003335
LC ebook record available at https://lccn.loc.gov/2020003336

Visit the Taylor & Francis Web site at
http://www.taylorandfrancis.com

and the CRC Press Web site at
http://www.crcpress.com

Contents

Preface

Understood as an industry concept recently proposed that encompasses the main technological innovations of the fields of automation, control and information technology applied to the manufacturing processes, Industry 4.0 is a real challenge for today's businesses. Indeed, from cyber-physical systems, Internet of Things and Internet Services, the production processes tend to become increasingly efficient, autonomous and customizable. The relationship with costumers is also more effective. In other words, the availability of real-time data enables organizations to offer more personalized products and customized solutions, which usually originates significantly higher margins than mass-manufacturing offers. Taking in mind these changes and implications in organizations, the present book, entitled *Industry 4.0 – Challenges, Trends, and Solutions in Management and Engineering,* looks to provide a support to academics, researchers and practitioners, as well as all to those who operate in the Management and Engineering fields and need to deal with the rapid pace of digitization in industry today. Becoming more active in Industry 4.0, these organizations see their benefits go far beyond extending the digital reach or selling new types of products and services. Industry 4.0 will establish these organizations, their employees as well as their total ecosystem of stakeholders, namely, suppliers, partners, distributors and customers as a complete interconnected and integrated digital network, linked to other networks around the world.

Taking into account this reality, interdisciplinary perspectives to further our knowledge and understanding of the developments in Industry 4.0 area are needed. Drawing on the recent developments and best practice, this book intends to examine the new advances in the subject of Industry 4.0, resulting from the last changes that have taken place and how they affect the management and engineering processes as well as the commitment and motivation of these organizations' workers.

Looking to provide discussion and the communication of the latest developments and thinking in what concerns the most recent research activity relating to this scientific field worldwide, with this book we look to highlight contributions that are exceptional in terms of theory, evidence and/or methodology and that significantly advance social scientific research on Industry 4.0 in management and engineering. It is designed to increase the knowledge and effectiveness of all those managers and engineers involved with Industry 4.0 in all kind of organizations and activity sectors.

The mission of this book is to provide channel of communication to disseminate the knowledge of new advances in *Industry 4.0 – Challenges, Trends, and Solutions in Management and Engineering,* and consequently the way how to manage competitive organizations between academics/researchers, managers and engineers.

Looking to promote research related to these new trends and developments in the field of management and engineering, the present book is divided into eight chapters. Chapter 1 covers "Emerging technologies and supply chain management – Maneuvering in current areas of tensions", while Chapter 2 discusses "Industry 4.0 creating a buzz in the western hemisphere: But watch out for China pulling into the fast lane". Chapter 3 focuses "Smart manufacturing based on Digital Twin technologies" at the same time Chapter 4 deals with "Industry 4.0 in small- and medium-sized enterprises – A literature review on research clusters". Chapter 5 contains information about "The role of industry 4.0 on the association between customers' and suppliers' involvement and performance improvement", while Chapter 6 highlights "Fundamental changes in the organizational processes: Industry 4.0 case study". Chapter 7 speaks about "Maturity level assessment for industry 4.0 integration into Lean Manufacturing", and finally Chapter 8 develops a reflection about "Human resource management in Industry 4.0 – A brief trip through literature".

Possible to be used by academics, researchers, human resources managers, managers, engineers and other professionals in related matters with Industry 4.0, the present book, entitled *Industry 4.0 – Challenges, Trends, and Solutions in Management and Engineering,* looks to:

- share knowledge and insights about Industry 4.0 on an international and transnational scale.
- keep at the forefront of innovative theories, models, processes and strategies, as well as the most recent research activities relating to industry 4.0.

Whether we are in the manager and engineer field, we need to know what is happening, on both national and international environments, to be able to understand and develop effective responses to meet all these new demands and challenges. This is why, today, the interest in *Industry 4.0 - Challenges, Trends, and Solutions in Management and Engineering* is evident for many types of organizations, namely, important institutes and universities all over the world.

The Editors acknowledge their gratitude to CRC Press/Taylor & Francis Group for this opportunity and for their professional support. Finally, we would like to thank to all chapter authors for their interest and availability to work on this project.

Carolina Machado
Braga, Portugal

J. Paulo Davim
Aveiro, Portugal

Editors

Carolina Machado has been teaching Human Resources Management subjects since 1989 at the University of Minho, Portugal. She has been an associate professor since 2004, with experience and research interest areas in the field of Human Resource Management, International Human Resource Management, Human Resource Management in SMEs, Training and Development, Emotional Intelligence, Management Change, Knowledge Management, and Management/HRM in the Digital Age. She is head of the Department of Management and head of the Human Resources Management Work Group at University of Minho, as well as chief editor of the *International Journal of Applied Management Sciences and Engineering* (*IJAMSE*).

J. Paulo Davim is a professor at the Department of Mechanical Engineering, University of Aveiro, Portugal. He has more than 30 years of teaching and research experience in Manufacturing, Materials, Mechanical, and Industrial Engineering, with special emphasis in Machining and Tribology. He has also interest in Management, Engineering Education, and Higher Education for Sustainability. He has worked as evaluator of projects for ERC (European Research Council) and other international research agencies.

Contributors

Shohin Aheleroff
The University of Auckland, Auckland, New Zealand

Ivan Erenda
TPV Group, Novo Mesto, Slovenia

João Carlos Espíndola Ferreira
Universidade Federal de Santa Catarina, Florianópolis, Brazil

Rui Pedro Freitas
School of Economics and Management, University of Minho, Portugal

Stefan Heng
Cooperative State University Mannheim, Mannheim, Germany

Erik Hofmann
University of St. Gallen, St. Gallen, Switzerland

Huiyue Huang
The University of Auckland, Auckland, New Zealand

Nicolai Kirstätter
University of St. Gallen, St. Gallen, Switzerland

Yuan Lin
The University of Auckland, Auckland, New Zealand

Yuqian Lu
The University of Auckland, Auckland, New Zealand

Carolina Feliciana Machado
School of Economics and Management, University of Minho, Portugal

Maja Mesko
University of Maribor and University of Primorska, Slovenia

Rogério Miorando
Universidade Federal de Santa Catarina, Florianópolis, Brazil

Julian M. Müller
Salzburg University of Applied Sciences, Puch bei Hallein, Austria

Marcos Malinverni Pagliosa
Universidade Federal de Santa Catarina, Florianópolis, Brazil

Jan Polzer
The University of Auckland, Auckland, New Zealand

Vasja Roblek
Faculty of Organisation Studies, Novo Mesto, Slovenia

Rapinder Sawhney
University of Tennessee, Knoxville, Tennessee, USA

Stefan Selensky
University of St. Gallen, St. Gallen, Switzerland

David Tomzik
The University of Auckland, Auckland, New Zealand

Guilherme Luz Tortorella
Universidade Federal de Santa Catarina, Florianópolis, Brazil

Alejandro Mac Cawley Vergara
Pontifícia Universidad Catolica, Santiago, Chile

Xun Xu
The University of Auckland, Auckland, New Zealand

Zexuan Zhu
The University of Auckland, Auckland, New Zealand

chapter one

Emerging technologies and supply chain management

Maneuvering in current areas of tensions

Erik Hofmann, Stefan Selensky, and Nicolai Kirstätter
University of St. Gallen, St. Gallen, Switzerland

Contents

1.1 Introduction

The far-reaching impact of the digitalization on business models and customer interactions has been widely discussed in the business press and scientific literature (Hu & Monahan, 2015, p. 96). What has been less discussed, however, is how digital technologies transform noncustomer-facing activities in general and supply chains (SC) in particular (Hu & Monahan, 2015, p. 96).

That indeed digital technologies hold vast potentials with regard to operations and supply chain management (SCM) was shown at the Hannover Messe in 2011, when Henning Kagermann, Wolf-Dieter Lukas, and Wolfgang Wahlster introduced the concept of Industry 4.0 (Kagermann, Lukas, & Wahlster, 2011). Since then, Industry 4.0 does not only refer to an increased automation of manufacturing systems, but also to an increasing conflation of the virtual and physical world based on cyber-physical systems and the so-called Internet of Things (Pfeiffer, 2017, p. 110). In other words, the transformation of noncustomer-facing activities in general and supply chain management in particular through digital technologies have come to the fore (Hofmann & Rüsch, 2017).

In order to fulfill this vision, decision makers need to redesign processes and integrate state-of-the-art digital technologies into their operations (Majeed & Rupasinghe, 2017, p. 25). Yet, solely digitalizing internal processes is insufficient. To reap the full benefits of adopting the Industry 4.0 concept and to stay competitive, companies have to leverage digital technologies when collaborating with their suppliers, partners, distributors, and, of course, customers. That is to say, companies have to digitalize their supply chain management. Despite the consensus among academics and practitioners on the great potential of digitalized supply chains, there are still plenty of "technical and nontechnical obstacles and challenges waiting" (Wu, Yue, Jin, & Yen, 2016, p. 411) to be tackled. These challenges are related to processes, technologies, management paradigms, systems, and capabilities of people (Handfield, 2016, p. 1). Addressing these challenges and potential solutions, however, is pivotal to realize the vision of a digitalized supply chain and to unlock the associated value potentials. Every category (Strategy & Business Model, Operating Model & Execution, Organization & Capabilities, IT and Data, as well as Technology) is affected by both SCM trends and the digitalization. As such, this chapter is intended to uncover digital supply chain challenges and potential solutions. Specifically, this chapter aims to answer the following research question: What areas of tensions are companies exposed to when digitalizing their supply chains? An area of tension refers to an option space for decision makers in the context of digitalizing the supply chain. The results' contribution of this chapter is threefold. First, the most pressing digital supply chain management challenges are identified. Second, those challenges are synthesized into a tension framework to provide a structured overview for decision makers and further research. Third, the tensions are described in detail and potential solutions are introduced.

For this purpose, the remainder of this chapter is structured as follows: first, the data and method is presented, before a digital supply chain management tension framework is introduced. This framework was developed based on the findings from the semi-structured interviews and the complementary structured literature review. Fourteen tensions within the categories "Strategy & Business Model", "Operating Model & Execution", "Organization", "Data & IT" and "Technology" are identified. Each tension is described in detail with a particular focus on the solutions for the two opposite challenges that compromise a tension. Finally, the implications of these findings are discussed and evaluated in light of the current scientific literature. Additionally, implications for supply chain management decision makers are given. Lastly, an overall conclusion of the chapter is given.

1.2 Data and method

This chapter follows a dual approach compromising a structured literature review and semi-structured interviews to answer the main research question of "What areas of tensions are companies exposed to when digitalizing their supply chains?" Qualitative research can provide insights into unexplored phenomena (Bettis, Gambardella, Helfat, & Mitchell, 2015, p. 637). Given the novelty of the topic and the purpose of this study, a qualitative research design is appropriate.

A dual approach is conducted to enhance the reliability and validity of the findings. The increase in both quality criteria will be achieved through triangulation of the data sets. Triangulating different kinds of data can potentially generate a "more complete, holistic, and contextual portrayal of the researched phenomenon" (Jick, 1979, p. 603). At first, a tension framework following the "Defined 10 Marketing Tensions" by Leeflang, Verhoef, Dahlström, and Freundt (2014, p. 4) was developed based on a structured literature review. In a second step, the framework was complemented by findings of the semi-structured interviews and enhanced with industry insights.

1.2.1 Tension framework

The empirical part of this chapter is based on the tension framework introduced by Leeflang et al. (2014, p. 4). The researchers developed the framework for an analysis on how digitalization impacts marketing with a specific focus on the challenges that companies encounter when going digital (Leeflang et al., 2014, p. 1).

The tensions help to structure the complex and rapidly changing environment SCM executives face. The framework helps to understand the changes and to develop a view on how to cope with them (Day, 2011, p. 183; Leeflang et al., 2014, p. 4). A tension and the associated challenges can be interpreted as either opportunity or threat but, in any case, require an answer from SCM executives (Leeflang et al., 2014, p. 4).

Therefore, the tension framework perfectly fits the research question and the objective of this chapter. The areas affected by the digitalization impact and SCM trends elaborated will be used as categories in the tension framework and thus drive the research process and data analysis. The categories are as follows:

1. Strategy & Business Model
2. Operating Model & Execution
3. Organization & Capabilities
4. Data & IT
5. Technology

After the basic framework of this chapter was introduced, the remaining paragraphs of this chapter are dedicated to the research process and data collection.

1.2.2 Research process

The literature documents and interview transcripts were analyzed using the category-centered approach by Mayring (2014). Content analysis in general was chosen because of its flexibility in terms of the research design (Harwood & Garry, 2003, pp. 479–480). The methodology provides the means to systematically and objectively describe as well as quantify specific phenomena and to draw realistic conclusions from the data (Bengtsson, 2016, pp. 9–10; Downe-Wamboldt, 1992, p. 314). Given the purpose of establishing a consolidated and comprehensive description of a phenomenon under investigation (Elo & Kyngas, 2008, pp. 109–110), the methodology is appropriate to answer the main research question of this study. In order to structure and carve out the challenges and tensions from the data, inductive content analysis according to Mayring (2014, p. 79 ff.) was executed. It is deemed appropriate for areas in which no previous research is available (Elo & Kyngas, 2008, p. 109). Within this economic procedure, the categories are established from the data itself and not from theoretical review (Mayring, 2014, p. 79). An important step is to define the "level or theme of categories to be developed" (Mayring, 2014, p. 80) beforehand by establishing a selection criterion. This deductive component is derived from the theoretical analysis of the subject and the research objective (Mayring, 2014, p. 81). For the chapter on hand, the category definition is based on the categories derived from the theoretical analysis and formulated as follows: Challenges in the digitalization of the SC in and around Strategy & Business Model, Operating Model & Execution, Organization, Data and IT, and Technology. The level of abstraction is concrete challenges and courses of action associated with digital initiatives in the SCM context (no general evaluations of the situation). The coding unit is defined as clear semantic unit in the text, whereas the context unit includes the entirety of the literature documents and interviewee transcripts (Mayring, 2014, p. 84). The challenges for SCM in the context of the digital era were identified by coding relevant text passages as well as by analyzing the frequencies as quantitative closing step (Mayring, 2014, p. 86). This approach ensures that the material relevant for the research question is analyzed appropriately and reflected holistically by the categories (Mayring, 2014, pp. 79–81). After the challenges were identified, they were matched with their antitheses in order to build the tensions. The benefits of the inductively category-driven content analysis procedure are "that it is strictly controlled methodologically and that the material is analyzed step-by-step" (Kohlbacher, 2006, p. 10).

Figure 1.1 shows the overall procedure undertaken by this chapter and highlights the relevant steps in the research process.

While the analysis procedure is reproducible with the research process depicted in the figure above, additional information regarding the sampling as well as the analysis and assignment of data to the categories is required.

1.2.2.1 Structured literature review

As mentioned above, a structured literature review was conducted as the first step in the procedure of establishing a tension framework for SCM in the era of digitalization. In line with Rowe (2014), the objective is to get a "good and reasonable coverage" (p. 246) instead of achieving comprehensiveness with respect to the economics and time constraint of this chapter. Following the arguments of Webster and Watson (2002), a review of existing literature "creates a firm foundation for advancing knowledge" (p. xiii), helps to "tackle an emerging issue" (p. xiv) and to develop a "conceptual model" (p. xiv). Therefore, the structured literature review is appropriate as a first step in developing the tension framework. The author has adopted a structured and systematic approach for selecting the appropriate papers based on methods inspired by literature about conducting reviews (Rowe, 2014; Tranfield, Denyer, & Smart, 2003; Webster & Watson, 2002) as well as other reviews (Fayezi, Zutshi, & O'Loughlin, 2016; Laursen & Svejvig, 2016).

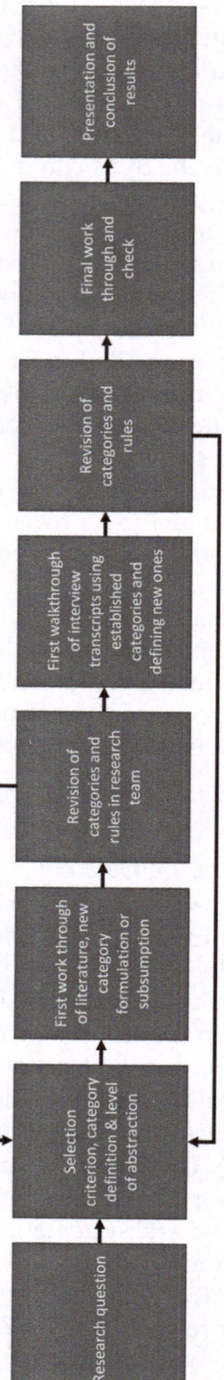

Figure 1.1 Research process. (Illustration based on Mayring, 2014.)

Table 1.1 Number of articles per category

Category	Number of articles
Strategy & Business Model	20
Operating Model & Execution	39
Organization	16
Data & IT	48
Technology	38
All categories	**161**

The approach comprises the following five steps: (1) planning and scoping, (2) conceptualizing, (3) searching, (4) screening and selecting, and (5) analyzing. Only English and German publications at the interface of digitalization and SCM were considered (1) before establishing the keywords to identify relevant literature (2). In a third step, a broad variety of papers such as peer-reviewed articles, consultancy reports, internet articles, and white papers were included in the web-based search given the novelty of the topic and revealed 58 articles (3). Those were then screened and selected based on their abstract, introduction, conclusion, and, where applicable, table of contents (4). Twenty-three articles were excluded because they were either not relevant to the topic of the chapter at all, too focused on a specific technology or element of the SC, or completely targeted at one company that left the author with 35 applicable papers to analyze (5).

Table 1.1 shows the number of articles per category. The total number of articles over all categories is larger than the number of articles included in the research as some articles cover multiple categories. The table shows that each category is covered by multiple articles and thus a broad research view per category is guaranteed. In summary, the articles per tension range from 5 (SC tasks) to 18 (SC equipment). Thereby, a sufficient amount of literature per tension is guaranteed. Text passages were allocated to the five main categories mentioned above and subcategories were established. The tensions per se were established by comparing the codes within and across the subcategories and main categories. Whenever codes were opening up a (contradictory) scope of action, a tension was established and thereafter elaborated with further codes. Second, not all passages coded were relevant for the formation of a tension and thus excluded in the course of the analysis. However, if an insufficient amount of codes (i.e., fewer than ten codes per tension and/or fewer than five codes per challenge) supported the tension, further analysis was dropped. After the literature coding, 19 challenges were established and afterward discussed and refined in the research team.

1.2.2.2 Semi-structured interviews

As outlined above, semi-structured interviews were the second source of data for this chapter. Due to the novelty of the topic, identifying potential interviewees experienced at the interface of digitalization and SCM proved difficult. Nevertheless, 22 potential interview candidates with backgrounds in SCM, digitalization, consulting, IT, and logistics were identified. Of this, 12 candidates responded with being generally available for an interview. Two of the 12 candidates, however, were unavailable during the research period. Therefore, a final number of ten interviews (45% of the total identified candidates) were conducted. Due to the heterogeneity of the interviewee's backgrounds and the novelty of the topic, a semi-structured interview approach was chosen. A pool of questions regarding the topics digitalization, SCM, and potential challenges and input from the literature review was compiled. Analyzing the transcripts, text passages were allocated to the five main categories and the subcategories, where applicable, established during the literature coding. New subcategories were defined where required and the above-stated criteria were met. Table 1.2 shows that the codes are quite equally distributed among the categories defined beforehand with one exception each to the top (Data & Technology) and to the bottom (Strategy & Business Model).

During the last two steps of the research process, the revision of categories and rules as well as the final work through and check, the initial tension framework defined from the literature review was finalized. In particular, tensions from the literature review were refined with codes from the interview analysis, dropped because of lacking SCM applicability or focus and added due to practical relevance. The detailed interpretation of the results will be done in following chapter.

1.3 Results

1.3.1 Tension framework

This chapter will detail and discuss the tension framework developed through the research process outlined above. The chapter is structured as follows: in the subsequent paragraphs, general remarks are introduced before the final overall framework is presented. Afterward, the identified digital SCM tensions are discussed in detail. It is important to bear in mind that first, as outlined above, each of the tensions can be regarded as opportunity or threat by businesses in the SC and that for each tension-specific characteristics like a company's position in the SC or industry have to be considered (Leeflang et al., 2014, p. 4). Also, the tensions do not exist in isolation, so that interdependencies have to be taken into account. With this limitation in mind,

Table 1.2 Interview coding and code distribution

Category	Interview partner 1	Interview partner 2	Interview partner 3	Interview partner 4	Interview partner 5	Interview partner 6	Interview partner 7	Interview partner 8	Interview partner 9	Interview partner 10	Total codes per category
Strategy & Business Model	8	6	7	11	1	8	3	10	18	8	**80**
Operating Model & Execution	16	15	11	6	2	12	6	17	13	11	**109**
Organization	14	5	7	10	0	3	1	30	22	10	**102**
Data & IT	21	17	22	8	1	11	7	28	15	16	**146**
Technology	6	7	8	7	2	12	5	27	12	11	**97**

Figure 1.2 shows the tension framework developed from the literature review and the interviews.

Overall, 14 digital SCM tensions were identified within the five main categories during the empirical analysis. Those tensions represent a synthesized overview of tensions that SCM functions experience when digitalizing the SC. The tensions might differ according to the industry, position in the SC, business model, and country. To gain a deeper understanding of the framework, the next subsections will continue with an in-depth discussion of each of the identified digital SCM tensions. For each tension, a short introduction to the topic will be given followed by a detailed explanation of the two challenges augmented by insights from literature and interviews.

1.3.2 *Digital supply chain management tensions*

1.3.2.1 *Strategy and business model tensions*

1.3.2.1.1 Tension: Business model – Adapt versus transform Digital technologies, as all other revolutionary technologies, concurrently threaten incumbent business models and open up opportunities that can be seized by new players (Vendrell-Herrero, Bustinza, Parry, & Georgantzis, 2017, p. 72). Even though business models are not an elementary part of SCM, the digitalization has the potential to profoundly change the way of how business is done within companies and the different functions along the SC. Increased client expectations (Interview partner 3) and new ways of client interaction (Kersten, Seiter, See, Hackius, & Maurer, 2017, p. 13) could trigger an adaptation or transformation of the existing business model.

Two developments are especially relevant for companies that operate along the SC: the increasing usage of platforms and servitization strategies. Originally used in the business-to-consumer (B2C) business, platforms are increasingly leveraged in business-to-business (B2B) for all kinds of transactions along the SC such as SC financing, purchasing, warehousing, and transportation (Kersten et al., 2017, p. 12; Schmidt et al., 2015, p. 7; The Economist Intelligence Unit Limited, 2017, p. 14; Zillmann, 2016, p. 28; Interview partner 4). Especially transportation platforms, such as InstaFreight or FreightHub, facilitate a more individualized logistics service for customers due to the access to and combination of multiple logistics providers (Kersten et al., 2017, p. 12). Platforms, however, can also have detrimental effects on the business model. On the one hand, product and service providers could lose end customer access, since intermediates become more relevant (Kersten et al., 2017, p. 13, Interview partner 4). On the other hand, the increased price transparency (due to direct comparison) on platforms raises the cost pressure on companies (Kersten et al., 2017, p. 38). In order to deal with the

Category		Tensions	Challenge	Description
Strategy & Business Model	1.	Business Model	Adapt vs. Transform	The digital transformation is forcing organizations and SC leaders to re-think their business models and roles within their SCs
	2.	SCM Strategy	Traditional vs. Digital	Given the increasing importance or data and technology for SCM, companies should consider adding those aspects to traditional SCM strategy elements
	3.	SCM Tasks	Operational vs. Strategic	The holistic overview and the increased data accumulation gives SCM the tools to make strategic tasks and decisions.
Operating Model & Execution	4.	Collaboration	Limited vs Expanded	Digital technologies facilitate the collaboration of SC partners but also increase the interdependencies and complicate risk management
	5.	Information Sharing	Join vs. Own Platform	Increasing requirements to share data with SC partners requires the decision to join an external platform or build up a proprietary one
	6.	Decision Making	Centralized vs. Decentralized	Whereas IOT and CPS enable decentral decision-making, SCM needs to decide which activities to centralize in order to achieve EoS and control
Organization	7.	Leadership	Treasurer vs. Collaborator	An overreliance on big data insights could stifle innovation and collaboration efforts in the SC
	8.	Talent Gap	Train vs. Hire	SCM and related departments need to decide on how to fill the talent gap in analytical and digital capabilities
IT & Data	9.	IT Infrastructure	License-and-install vs. XAAS	A powerful IT is a competitive advantage in the era of digitization and companies need to weigh the benefits and drawbacks of switching to XAAS.
	10.	Big Data Analytics / AI	Prescriptive vs. Predictive	SC leaders need to decide how much power and decision-rights they want to give to artificial intelligence
	11.	Blockchain	Wait-and-see vs. Orchestrate	Companies have to weigh Blockchain's advantages & applicability and lacking global technological standards, governance structures & harmonized national laws
Technology	12.	SC Features	Identify vs. Instrument	The integration of today's technologies enables not only the identification of objects in the SC but also the instrumentation of the SC to establish holistic transparency
	13.	Variable Material Flows	Supportive Technology vs. Robotic	In order to make non-standardized processes more efficient, companies can choose between supportive technologies and robotic solutions
	14.	Additive Manufacturing	Niche vs. Integrated Usage	Additive manufacturing can help to deal with some of today's more prevalent SC challenges and become an integral part of SC processes

Figure 1.2 Tension framework.

increasing usage of platforms, companies need to adapt their current business model. Opportunities include the listing on those platforms, the development of an own platform, or the investment into a start-up or existing platform provider (Interview partner 4).

The second development is the adoption of servitization strategies through which companies can generate additional value, sales, and profits (Vendrell-Herrero et al., 2017, p. 317). The increasing demand for individualized offerings constitutes a great challenge for companies. Offering client-tailored value-added services is an opportunity to increase customer satisfaction, customer retention, and differentiation in the competitive landscape (Kersten et al., 2017, p. 39; Zillmann, 2016, p. 28). Besides monitoring start-ups and cooperating with SC partners (Kersten et al., 2017, p. 43), data plays a pivotal role in the development of service offerings as Interview partner 8 highlights.

In sum, companies need to revise their business models to leverage the full potential of smart technologies as well as data analytics and to determine the required resources and capabilities to provide the services for their customers (Wu et al., 2016, p. 410). SCM plays a pivotal role in designing, running, and improving the processes to deliver new service offerings.

The more radical response is a transformation of the business model. The analysis of the literature and the interviews unveiled two reasons to transform the current business model. First, digital technologies endanger incumbent operations. Additive manufacturing technologies, for instance, could cause revolutionary changes in the logistics industry as the conventional value adding process of logistics providers, the transport of (manufactured) goods, could be significantly impacted (Thomas, Kammler, Zobel, Sossna, & Zarvic, 2016, p. 59). Consequently, logistics providers have to consider opportunities to adapt or even better transform their business model in the era of digital delivery and on-spot product manufacturing. Second, companies use the increasing amount of data to conduct a transformation toward a more usage-oriented business model (Kache & Seuring, 2017, p. 23). GE Aviation, for instance, changed from an engine manufacturer and vendor to a company selling flight hours (Mussomeli, Gish, & Laaper, 2016, p. 10). With the help of sensors attached to aircrafts, GE Aviation is able to gather and analyze sufficient data to decrease downtime and maintenance expenses and thereby increase the value for the customer (Mussomeli et al., 2016, p. 10). The offering of tailor-made solutions is a response to the increasing (individualized) demand of customers. Data plays a core role in designing, delivering, and improving such new business models.

However, before an adaptation or transformation is conducted, the development of a thorough vision and the specification of the digital business model is an important prerequisite to quickly and successfully adopt the business model to the changed environment (Kersten et al., 2017, p. 56).

1.3.2.1.2 Tension: Supply chain strategy – Traditional versus digital SC strategies have historically been focused on cost reduction, agile responses, or a hybrid of both (Madhani, 2017, p. 40). With the onset of the digitalization, however, data and technology have increasingly permeated all aspects of SCM. This raises the question if organizations should formulate specific strategic objectives regarding data and technology in their SC strategies.

Lowering or containing costs is still seen as major objective of SC managers (Butner, 2010, p. 23; The Economist Intelligence Unit Limited, 2017, p. 11). Most of the interview partners also highlight the importance of focusing on rather traditional strategic objectives and making data and technology more of an implicit part of those objectives. Interview partner 3, for instance, outlines that "every Industry 4.0 use case or every new technology, which is introduced, needs to influence time, quality or costs somehow positively". In the opinion of Interview partner 9, data and technology will become more important for the SC strategy but both have no "self-purpose" and the strategy will "not be about supply chain data and technology, but the utilization of data and technology to improve the supply chain".

The determination of a strategy includes making choices and thus trade-offs. By focusing on too many objectives simultaneously, original SC objectives could get blurred. Consequently, data and technology are rather seen as facilitators that help to realize traditional strategic SC objectives through targeted initiatives and pilots (Interview partners 4 and 8).

The traditional focus, however, poses the danger of neglecting the relevance of data and technology in the SC organization. By making data and technology an explicit part of the SC strategy, visibility can be raised and accountability improved. Interview partner 2 outlines that strategic objectives in his organization include the development of a global business intelligence platform and a harmonized IT infrastructure. Digital collaboration, data-driven decision-making, and predictive maintenance are strategic objectives in the company of Interview partner 10. In the company of Interview partner 8, the importance of data and technology is considered so important that a digitalized SC vision was developed. Within this vision, the first objective is to create complete connectivity, that is, every internal and external asset will be "assigned an IP address in the Internet of Things, on our platform" (Interview partner 8). Clear digital objectives in the SC strategy are the basis for long-term value generation and measurement, thereby supporting the digital transformation of the organization (Raab & Griffin-Cryan, 2011, p. 9). Additionally, pooling single initiatives in an overall strategic direction helps to create synergies.

The result is a digital SC strategy that gives clear guidance on how to be successful in an increasingly digital environment.

The digital strategy, however, needs to be aligned with the overall digitalization strategy of the company (Raab & Griffin-Cryan, 2011, p. 9).

1.3.2.1.3 Tension: Supply chain tasks – Operational versus strategic
SCM tasks are principally perceived as operational in nature or as enabler of the overall strategy and the business model. Cooper et al. (1997), for instance, defined SCM "as an integrative philosophy to manage the total flow of a channel from earliest supplier of raw materials to the ultimate customer, and beyond, including the disposal process" (p. 68). Also, Eßig, Hofmann, and Stölzle (2013) adopted a rather operational perspective in their definition of SCM as "the cooperative coordination of material, information and financial flows in cooperate networks" even though they highlighted the strategic aspect of "realizing competitive advantages at the end user's level and hence increasing profitability of the overall network" (p. 41). The two examples show the directions of impact.

The digitalization provides the SC organization with multiple tools to take on a more strategic role in the overall company. However, SCM should not lose track of its operational responsibilities. Hence, companies must decide whether or not to grant SCM more strategic tasks.

Traditionally, SC tasks have been divided into planning and execution with execution being subdivided into sourcing, making, and delivering, or a variation thereof (Mussomeli et al., 2016, p. 6). Those operational tasks are in the context of the digitalization still highly important as outlined by Schmidt et al. (2015) and as the interviews conducted for this chapter show. Interview partner 7, for instance, thinks that the digitalization will not have an impact on the "superordinate, normative thinking in SCM", but argues that specific operational tasks could change "with robotics solutions or virtual reality". Interestingly, some interviewees expect a stronger collaboration on the operational level as consequence of the digitalization. Interview partner 10 believes that the tasks and the role of SCM will be strengthened because "the holistic consideration of the units in SCM and then also a SCM across the SC is not realized yet". This is, however, necessary to realize the "actual notion and philosophy of SCM, that you can observe end-to-end" and reap the full benefits that are described by scientists and scholars. Interview partner 6 argues in the same direction and in his opinion, the dissolved functions within a company will close ranks. In summary, it can be observed that operational tasks remain important, that the digitalization might enable SCM to give input on selective strategic decisions, and finally, the emergence of a SCM function through a closer collaboration among the operational functions is a likely scenario.

Another option, however, is to give SCM far-reaching strategic influence. By this, the perspective on SCM could be transformed from a cost to a differentiation perspective thereby delivering the foundation for a competitive edge (Jocher, 2017). The three areas surfaced in which SCM can take over more strategic tasks are strategic decision-making and planning, revenue creation, and technology access.

First, the digitalization provides the basis for SCM to take on a pivotal role in strategic planning and decision-making (Mussomeli et al., 2016, p. 2). The constant data collection along the chain provides a comprehensive source of insights that can be leveraged for other business units and decision-making (Butner, 2010, p. 26). SCM can be a valuable input for strategic planning and decision-making with regard to three specific SCM trends risk/interruptions, sustainability, and regulations (Kersten et al., 2017, p. 19). An increasingly important trend is environmentally, socially, and ethically sustainable operations along the SC (Kersten et al., 2017, p. 20). SCM can be a valuable input on how to redesign the SC to reduce the carbon footprint. This not only represents a valuable tool for customer communication, but also potentially impacts the overall strategy and business model. The increasing demand for transparency regarding sourcing and sustainable behavior can be regarded as the strategic task of SCM (The Economist Intelligence Unit Limited, 2017, p. 4). Finally, as Interview partner 2 outlines, SCM can be of strategic relevance in the regulatory environment, for example, fulfilling local laws with regard to medical devices. Moreover, SCM can be a source for additional revenue. In the following, a couple of specific strategic areas are outlined. This overview, however, is not complete. Predictive analytics will open up opportunities for companies to turn products into "products-as-a-service" (Padmos et al., 2016, p. 11) as SCs are able to deal with customer needs in real time. A typical example is predictive maintenance. Besides adding services to the portfolio, the digital SC enables companies to offer smart products and smart logistics and highly customized products and experiences to its customers, thereby fulfilling an increasing demand for mass customization (Hanifan, Crosnier, & Timmermans, 2016, p. 4). Also, SCM professionals can have a positive impact on the revenue by suggesting new and providing quicker access to markets (Mussomeli et al., 2016, p. 2). Taking over more strategic tasks however requires a shift in thinking. SC success needs to be measured not only in terms of costs and cost drivers, but also in terms of revenue and profitability (Benton, 2017). Lastly, SCM can also be assigned with the responsibility to give the organization access to strategically important Industry 4.0 technologies as well as the respective resources (Stölzle, Hofmann, & Mathauer, 2018, p. 92). Due to the daily interaction with outsiders in both directions of the SC, SCM employees get a feeling of current trends and technologies

(Interview partner 9). In consequence, SC executives will increasingly be responsible to put the SC into a position to make valuable strategic contributions to the progress of the company.

1.3.2.2 Operating model and execution tensions

1.3.2.2.1 Tension: Collaboration – Limited versus expanded Today's SCs are often a series of separate and siloed steps inside and across companies (Schrauf & Berttram, 2016, p. 4). Given the augmented competition, speed, and complexity, companies increasingly connect and collaborate to improve their competitive position in the market (Kersten et al., 2017, p. 21). The digitalization gives companies the tools to effectively collaborate across a multitude of dimensions (Schrauf & Berttram, 2016, p. 4), while simultaneously changing the mode of collaboration and composition of players within the SC (Howells, 2016; Kersten et al., 2017, p. 39). Collaboration can be increased along two dimensions: vertical and horizontal. The advantages of increased collaboration are mainly rooted in higher flexibility and faster adaptability as well as the joint optimization of the overall network (Kersten et al., 2017, p. 21; Interview partner 6). The option spectrum shows that the opportunities of collaboration are multifaceted and cannot be synthesized into a bipolar measure easily. In the following, limited and expanded collaborations are distinguished as the two broad streams of collaboration opportunities. Whereas limited collaboration refers to the typical operational SC tasks, expanded collaboration goes beyond operational and vertical collaboration.

Selective cooperation among partners in the SC is already happening today on a wide scale. Typical SC tasks that can be tackled together include product and process improvements as well as capacity and need planning, alignment, and co-developments (Alicke, Daniel Rexhausen, & Seyfert, 2017; Interview partners 6, 9, and 10). Besides operational topics, companies can co-develop products. A strategic co-development partner can prove beneficial for noncore product developments (Interview partner 6). However, the question arises why companies do not increase their collaboration efforts if the technologies exist and the benefits are proven. The analysis revealed three main reasons: risk, competition, and technological barriers. Increased globalization, SC interdependence, and intertwined business activities among SC partners and clients elevate the risk exposure and the difficulty to manage risk adequately (Butner, 2010, p. 25; Schmidt et al., 2015, p. 3). The fear of competition and the exploitation of information sharing is another reason that prevents companies from augmenting cooperation (Interview partner 9). Lastly, technological barriers could hinder companies to align business activities and collaborate. Inadequate IT, lacking interfaces, and missing processes impede cross-company management of the SC (Butner, 2010, p. 24).

Despite such reservations, the potential benefits of an augmented collaboration could incentivize companies to extend their common activities. The expanded collaboration can be subdivided into horizontal and vertical activities. On the horizontal axis, crowdsourcing and open innovation could improve and accelerate the design process (Mussomeli et al., 2016, p. 11). Additionally, universities and research organizations are a valuable source for product and process innovation (Jocher, 2017; Papert & Pflaum, 2017, p. 179). Finally, given the increased relevance of software and digital technologies, IT companies become a valuable source to co-develop and implement a new vision (Hanifan et al., 2016, p. 6; Interview partner 8). Interview partner 8 points out that his management is considering implementing their new vision of a digitalized SC in collaboration with a leading software company.

On the vertical axis, companies can choose to expand their collaboration on the operational, tactical and strategic level (Schrauf & Berttram, 2016, p. 14). Integrated planning, collaborative decision-making, and collaborative execution show several ways of increasing the collaboration efforts (Schrauf & Berttram, 2016, p. 6; Wu et al., 2016, p. 400). Such integration enables SCs to react rapidly and coordinately to problems by capitalizing on a broad and integrated (real time) data basis (Butner, 2010, p. 29). The vision of the expanded collaboration across levels are end-to-end processes, that is, processes that are connected from the supplier to the end customers (Raab & Griffin-Cryan, 2011, p. 7; Schrauf & Berttram, 2016, p. 27). This provides the basis for the ultimate process design target: completely automated processes without manual intervention (Raab & Griffin-Cryan, 2011, p. 27). Technologically, integration platforms in addition to the information platforms provide the infrastructure for collaborative planning, day-to-day operations, and scenario analysis (Schrauf & Berttram, 2016, p. 14). However, as outlined above, increased integration can cause an elevated risk exposure and complexity. Hence, a coordinated approach to risk analysis and an effective risk management is inevitable.

Asset sharing is an increasingly popular collaboration type on the operational level (Alicke et al., 2017; Howells, 2016; Mussomeli et al., 2016, p. 11; Schmidt et al., 2015, p. 3). Companies can increase asset utilization and transport flexibility as well as target new markets by sharing capacities (e.g., trucks, containers, wagons) with other stakeholders via electronic platforms (Alicke et al., 2017; Schmidt et al., 2015, p. 3). Besides optimizing intentions, companies also share assets with companies that do not possess or cannot afford more advanced capacities such as telematics modules. Interview partner 7 explains that suppliers of the company he is working for are equipped with telematics modules to support and monitor the transportation of perishable products. Lastly, companies can collaborate in strategic optimization projects (Schrauf & Berttram, 2016, p. 15). This area, however, seems to be quite untouched based on the analysis of literature and interviews. This is most likely due to the sensitivity of the information involved.

In order to effectively and successfully manage increased collaboration and integration, structured collaboration models, key performance indicators, and an appropriate governance structure are key prerequisites (Hu & Monahan, 2015, p. 96; Kache & Seuring, 2017, p. 27; Schrauf & Berttram, 2016, p. 28). Such structures support the mediation and resolution of conflicts and provide the basis for compliance to the predefined collaboration rules (Kache & Seuring, 2017, p. 27).

1.3.2.2.2 Tension: Information sharing – Join versus own platform
Driven by the end customers, end-to-end transparency across all SC stakeholders is not only an operational but also increasingly a strategic imperative for companies (Jacques, 2017, p. 20). Transparency refers to the "traceability from the starting product to the end customer" (Kersten et al., 2017, p. 32). Interview partner 10 refers to transparency as "not only the traceability of single articles but actual transparency of [...] who are the stakeholders, not only the suppliers but also the sub-suppliers". To create a holistic overview of the SC, gathered data needs to be shared with other stakeholders in the SC. Cloud-based information platforms are an opportunity to make data centrally available.

SC cloud platforms are joint infrastructures between a company, its suppliers, and partners as well as clients. Multiple stakeholders can share general and real-time data directly with each other. The scope can be increased to joint planning or even collaboration solutions (Alicke et al., 2017). Important requirements for such platforms are adequate interfaces as well as standardized data formats (Zillmann, 2016, p. 7).

Information exchange within the network is pivotal for the digital transformation of the SC, and central platforms enable a new level of collaboration (Alicke et al., 2017; Zillmann, 2016, p. 13). In order to realize increased visibility and transparency across the SC, companies can choose between setting up their own platform or choosing and joining an external solution. The decision should be based on a qualitative and quantitative evaluation that includes, among others, cost, system security, data format types, and the ease to integrate clients and suppliers (Kersten et al., 2017, p. 41).

Information sharing benefits on a central platform are far-reaching. A prerequisite for the optimization of the SC is the availability of real-time end-to-end information (Kache & Seuring, 2017, p. 23). With the help of central information platforms, decisions and SC steering are conducted on the same data basis and the increased multitier visibility helps to improve agility and efficiency across the entire network (Kache & Seuring, 2017, p. 24). The multitier visibility is an important topic as companies often do not specifically know their sub-suppliers and associated risks. With full visibility, companies can react to potential disruptions in the SC in real time. Additionally, increased transparency helps companies to spot critical bottlenecks, such as scarce resources, and to develop risk management mechanisms and diversify sourcing (Interview partner 9).

Generally speaking, companies can choose between joining an established platform or setting up their own. Often, however, companies have no choice but joining an established infrastructure of a big player. This step needs to be well thought through. First of all, if a bigger company (e.g., an original equipment manufacturer (OEM) in the automotive industry) is the focal point of the network and owns the information platform, there is the risk that the OEM optimizes its own operations and inventories instead of optimizing the overall network (Interview partner 3). Second, security standards may vary between operators and are difficult to verify for externals. Third, the company possesses the data sovereignty and manages all information flows. Especially, the latter data-related issue could keep companies from joining externally managed platforms (Interview partner 3).

An impartial entity could produce relief. The Fraunhofer Institute, in collaboration with the German Bundesministerium für Bildung und Forschung and leading enterprises, develops the so called "Industrial Data Space" (Zillmann, 2016, p. 19). As the data room is operated by an impartial entity, the threat of optimizing single companies is minimized and data sovereignty is maintained. If these trust issues will not be wiped out by impartial intermediates, reservations regarding data sharing will maintain and the usage of such platforms will remain a niche phenomenon (Interview partner 8). Another advantage of an impartial entity owing the platform is the setting of cross-sector (data) standards (Theisinger, Specht, & Hasenstab, 2015, p. 6; Zillmann, 2016, p. 18). If companies develop their own platforms, data standards differ and hence complexity and costs of joining increase for suppliers and partners. As of yet, such standards have not been rolled out industry-wide, but first efforts seem promising: the reference architecture model RAMI 4.0 is being developed as basis for digitalized industrial SCs and the electronic airway bill (e-AWB) is an industry-wide initiative to fasten the contract between an airline and the freight forwarder (International Air Transport Association, 2017; Zillmann, 2016, p. 18).

All stakeholders in an ecosystem can be connected without media disruptions using the highest security standards on such a cloud infrastructure (Zillmann, 2016, p. 18). Nonetheless, the major challenge for providers is to develop a platform that fulfills all relevant stakeholder expectations in the areas of data security, data linkage, data exchange, functionality, and availability (Zillmann, 2016, p. 18).

Consequently, the development of a proprietary information platform could be the solution to customize the functionalities according to the specific needs of the company and its SCs. Besides information functions, operators could give suppliers and customers self-service functions such as the change of reference data or direct job management (Kersten et al., 2017, p. 31). With a proprietary platform, companies are in absolute control over their data and can assign specific access rights to stakeholders in the SC. The development of a proprietary platform however faces three specific challenges: technological infrastructure changes, monetary investments, and partner involvement. Even though it might sound trivial, the integration of a data platform into existing software solutions can prove difficult (Interview partner 4) and is often associated with high investments. Additionally, as touched upon above, other players are required to join the platform. They need to be convinced of the data sharing on one's proprietary platform.

This requires a general change of mind among SC stakeholders. Companies are still reluctant to share operational data not to mention financial and revenue data (Interview partner 8), although it can bring fundamental benefits (Hofmann, 2007). Nonetheless, in order to reap the benefits of the vast amount of data, sharing is essential among the relevant stakeholders (Kersten et al., 2017, p. 36).

Companies that plan to invest into or join a platform also have to consider the risks and challenges involved. Besides the distribution of sensitive information, cyber security issues, the quality of shared data, and lacking data semantics need to be taken into consideration (Zillmann, 2016, p. 16).

1.3.2.2.3 Tension: Decision-making – Centralized versus decentralized

In the context of SC digitalization, companies are facing the challenge of increasing complexity, individualization, and demand fluctuations while simultaneously increasing efficiency (Kersten et al., 2017, p. 19). The decentralization of the operating model and the organization could prove as a productive approach to become more flexible and to cope with this trade-off. The decentralization can be separated into decentral physical entities and decentral decision-making (Kersten et al., 2017, p. 21). One example of decentralized physical structures is the integration of additive manufacturing technologies into the value creation process. More and smaller warehouses for a faster delivery are another example of decentral physical entities (Kersten et al., 2017, p. 21). Concurrently, a combination of technologies (e.g., CPS, IoT, and artificial intelligence) facilitates decentral decision-making and process steering.

In contrast, the implementation of a central governance structure might be rooted in the demand for higher efficiency and the wish for a holistic SC supervision and control. An increasingly popular approach for central SC supervision and decision-making is the so called "SC control tower" (Gstettner, Roesgen, Ganeriwalla, Kotlik, & Walter, 2016; Schrauf

& Berttram, 2016, p. 10). It provides decision makers with the full visibility of a company's SC and the relevant information to make decisions (Schrauf & Berttram, 2016, p. 19). Usually, those towers are staffed with cross-functional teams that gather SC data about interruptions, demand fluctuations, capacities, and inventories among others in real time, use algorithms to analyze the data and derive subsequent actions (Gstettner et al., 2016). With the introduction of CPSs to the SC, however, many decisions can be delegated to local entities. Interconnection puts CPSs into the position to make large-scale optimization decisions and to increase responsiveness to upcoming changes. Thereby, fully automated process flows and the integration of material, information, and financial flows is enabled. The interviews indicated that decentralized decision-making with the help of CPSs is still in its infancies (Interview partners 1, 3, 4, 6, 7, and 8). Interview partner 3 explains that many OEMs move decentral and automated decision-making forward in terms of pilots and proof-of-concepts (see Tension XII for the example of a Bavarian OEM).

Even though pilots are underway, Interview partners 1, 3, and 6 highlight that many decisions are still made centrally. Decision-making based on CPSs requires powerful information technologies and integrated data flows that are analyzed in real time. This might be a root cause of why many companies still struggle to decentralize decision-making and process steering.

1.3.2.3 Organizational tensions

1.3.2.3.1 Tension: Leadership – Treasurer versus collaborator The eclectic changes that the SC digitalization causes and the multitude of external trends, such as increased cost pressure, complexity, and risk exposure (Kersten et al., 2017, p. 19), impact the requirements for leadership management and leadership competencies. The introduction of new technologies and the increasing usage of Big Data in the context of digitalization require a new skillset and approach to leadership (Kersten et al., 2017, p. 44). Basic knowledge of relevant technologies and programming languages, the firm application of IT, and data analytics competencies are increasingly important skills for SC managers (Kersten et al., 2017, p. 65). Additionally, the adaptation of the leadership style could be pivotal for the successful digitalization of the SC (Kersten et al., 2017, p. 44). Generally speaking, two leadership style manifestations can be distinguished: a rather numbers-based treasurer style or a more human-focused collaborator style.

The digitalization gives decision makers more choices and higher-precision data to optimize global networks on a real-time basis (Butner, 2010, p. 24). Intelligent systems synthesize and analyze data based on potential constraints, present alternatives, and propose decision options with ideal outcomes (Butner, 2010, p. 27). Whereas dashboards are often based on backward-looking operational KPIs that have not been adapted in the last 20 years (Interview partner 1), the increasing data basis enables real-time KPIs that can be regularly adapted to changing business needs and environments (Interview partner 9). A numbers-based approach also helps to manage two relevant exogenous trends that currently affect SCM: cost pressure and increased risk exposure (Kersten et al., 2017, p. 19). Executives see an increasing need for treasurer expertise for the SC of the future. The value added of a treasurer's expertise derives from improving risk management, monitoring cash and liquidity management, streamlining financial processes, and increasing efficiency (The Economist Intelligence Unit Limited, 2017, p. 22).

However, since SCM encompasses a multitude of functions, companies, and decision makers, collaboration skills are also important for SC business leaders (Butner, 2010, p. 22). A data-driven leadership with an overreliance on information could impede the integration of relevant stakeholders and joint planning, risk mitigation, and execution (Butner, 2010, p. 29; Kache & Seuring, 2017, p. 21). Unlocking all benefits of a digitalized and collaborative SC also requires innovative thinking and looking at the overall network (Hu & Monahan, 2015, p. 95; Zillmann, 2016, p. 27). An overarching transformational SC leadership style is thus another alternative to manage the SC in the context of the digital transformation (Stölzle, Hofmann, & Oettmeier, 2017, p. 70). The individual recognition of employees, strong communication skills, and an inspiring motivation can be beneficial in times of major structural and operational changes (Kersten et al., 2017, p. 46; Stölzle et al., 2017, p. 46). Especially during times of accelerated change, a human-focused leadership is essential to release fears by listening to employees, convincing them of the future vision, and eventually decreasing the resistance to technological change (Kersten et al., 2017, p. 45; Interview partners 8, and 9).

The digital transformation can also be supported by a change in culture. Interview partner 9, for instance, believes that companies "are less hierarchical as they were previously, because the time has become more fast-paced, and it is less about to push through ideas against resistance". The leadership needs to be open toward the many opportunities that are available, and more flexible in choosing and switching between ideas, something that Interview partner 9 denotes as "keeping several balls in the air, and being open to continuously change priorities and opinions".

As planning cycles shorten and decisions have to be made faster, experimenting becomes an integral part of in-house operations. Consequently, a positive handling of mistakes or the establishment of a fast failure culture will facilitate the propensity to experiment. An environment with less fear to fail and pressure to justify might ultimately yield greater innovation (Mussomeli et al., 2016, p. 15).

Leadership in the time of digitalization is, according to Interview partner 1, also a generational question. Change in leadership and culture can build even more momentum as soon as "the digital generation gets into the management".

1.3.2.3.2 Tension: Talent gap – Train versus hire Sufficient and appropriate capabilities are a key enabler of a successful SC digitalization. Data analytics and management of systems, historically tasks of IT departments, are increasingly required capabilities for SC employees (The Economist Intelligence Unit Limited, 2017, p. 22). The lack of data scientists who can analyze data from different sources and extract patterns as well as coherences and the shortage of computer scientists who can program and work with algorithms and ensuing artificial intelligence solutions is not only a major challenge for SC functions but also organizations in general (Kersten et al., 2017, p. 44). Besides data analytics capabilities, SC employees need to be able to derive and structure appropriate actions based on data insights. One reason for the talent gap is the fact that today's required digital capabilities go beyond the knowledge of most computer scientists that were educated 10–20 years ago (Interview partner 9). This lack of adequate resources impedes the digital transformation of most SCs (Kache & Seuring, 2017, p. 26; Kersten et al., 2017, p. 20; Stölzle et al., 2017, p. 55). Organizations have two levers to handle the resource shortage: train existing employees or hire new people. At first, however, an analysis of the prospective required capabilities needs to be conducted (Zillmann, 2016, p. 27). The basis for that is the realistic evaluation of existing capabilities and the rigorous pointing out of gaps and lacking capabilities (Zillmann, 2016, p. 27).

Choosing to train employees emerges through three reasons: First, experts are available within different departments of the organization but have not been discovered yet (Kersten et al., 2017, p. 52). Interview partner 8 outlines that even though the SC function in his organization is lacking experts on Artificial Intelligence (AI), they can be found in the autonomous driving business unit. In order to discover relevant resources within an organization, employee capabilities need to be identified across jobs and business units (Kersten et al., 2017, p. 52). Second, the hiring of new resources could cause layoffs among the existing staff (Interview partner 8). And third, hiring new personnel is too expensive (Interview partner 10).

Companies can choose from a multitude of options to train and integrate required digital resources. Besides the setup of internal learning laboratories, intergenerational teams, and intrapreneurship, companies can also collaborate with universities and research institutes or participate in funding initiatives (Kersten et al., 2017, pp. 51–53).

However, the capabilities upgrade faces a multitude of internal and external hurdles that need to be overcome. Internal hurdles include the lack of financial, personnel, and temporal resources; internal training concepts; and employee willingness (Kersten et al., 2017, p. 47). Among external hurdles are insufficient offer quality and the danger of people outflow after training (Kersten et al., 2017, p. 47). Besides training investments into professional expertise, leadership and methodological competences need to be trained as well (Fischer, 2017). This is important in order to initiate a cultural change that is required for the SC digitalization (Schmidt et al., 2015, p. 10; Schrauf & Berttram, 2016, pp. 27–28).

Often, however, new resources are either not available internally or an in-house build-up would take too much time. Additionally, new and external people might facilitate and accelerate the cultural change (Interview partner 8). Companies have a multitude of external opportunities to close the talent gap ranging from short-time purchase of consulting or data analytics services, over the hiring of full-time employees to the acquisition of a start-up or an established company.

An increasing number of providers offer data science as a service (Kersten et al., 2017, p. 52). This gives companies the leeway and flexibility to request data analytics services on a demand basis without committing to additional resources. Besides data as a service, companies draw on service providers to build up IT systems and teach analytics software.

If data analytics and digital technologies become an important part of the business operations, the full-time hiring of specialists is reasonable. Externals can be advantageous as "a new and fresh mindset" (Interview partner 8) can be internalized that could initiate a change of mind within the company. Often the hiring of a new resource is the first step to get the required know-how into the organization (Interview partner 9). Additionally, if a fundamentally new digital SC vision is developed, a new team can be advantageous for a potential transformation process. Given the shortage of adequate talents, companies have to offer above-average salaries and other related benefits to remain attractive (Interview partner 8). The most vigorous response to the talent gap is acquiring a company that is equipped with the respective expertise.

1.3.2.4 IT and data tensions

1.3.2.4.1 Tension: IT infrastructure – License and install versus XaaS An appropriate IT infrastructure is a key success factor for the SC digitalization. Given the integrative nature of companies and functions within a SC, technological integration plays a pivotal role in the realization of the digital SC. Defining common standards and interfaces across the SC is paramount given the importance of cooperation in the SC (Kache & Seuring, 2017, p. 28). However, before a cross-company integration can be implemented, systems within companies need to be aligned. Additionally, data sharing and Big Data analytics can only be utilized effectively with standardized and integrated IT structures (Interview partner 3).

In general, companies either create interfaces among their current systems (Butner, 2010, p. 27) or leverage cloud computing (CC) solutions that provide standardized interfaces (Kersten et al., 2017, p. 28).

Given their closed nature and limited cross-company functionalities, conventional enterprise resource planning (ERP) systems are often insufficient for today's SCs (Jede & Teuteberg, 2015, p. 872). In contrast, CC solutions offer substantial advantages for SC managers to meet today's SC conditions.

The key advantage of CC systems is the usage of common standards for data formats, interfaces, processes, and security mechanisms across companies in the SC (Zillmann, 2016, p. 20). The creation of interorganizational interfaces, the feature to easily change current linkages and end-to-end data exchange are paramount for the optimization of the SC (Jede & Teuteberg, 2015, p. 890). The alteration of existing linkage is particularly beneficial given the decentralized nature of global SCs and the development toward supply networks that quickly change their compositions depending on end customer demand.

Given that a vast amount of data will come from objects (e.g., augmented reality glasses and hand scanners), those objects need to be integrated into the ecosystem as well. The fragmented IT landscapes with legacy systems in many corporations pose a threat to IT and SC effectiveness (Kache & Seuring, 2017, p. 28). CC solutions deliver the means of interoperability and compatibility so that systems, software applications, machines, and eventually companies can be integrated efficiently.

Additionally, total cost of ownership of cloud-based solutions has been proven to be lower than traditional license and install packages (Jede & Teuteberg, 2015, p. 885; Kersten et al., 2017, p. 52). First, XaaS solutions that are sold on a "pay-per-use" pricing model can be purchased on a demand basis (depending on the resource requirements) (Kersten et al., 2017, p. 52). Second, investment costs are minimized and updates as well as repair costs are usually included in the price (Kersten et al., 2017, p. 52). The minimization of investment costs is especially beneficial to small- and medium-sized companies for which the IT integration has been impeded by high one-time investment costs into traditional software solutions (Schurk, Harth, Herrmann, Rosina, & Scheid, 2015, p. 12). The third benefit of CC applications is the value and performance increase of IT systems (Jede & Teuteberg, 2015, p. 886). In particular, add-on services, constant updates into the software and hardware, and service bundle offerings are likely to increase performance and drive the value for SCM (Jede & Teuteberg, 2015, p. 886).

In sum, CC offers three main benefits: higher interoperability, lower cost of total ownership, and increased IT performance and value.

Nonetheless, CC also has some downsides that need to be considered carefully. Key considerations are cyber security and system stability risks. CC providers often concentrate a high amount of system nodes as well as data and are thus likely to attract hackers (Jede & Teuteberg, 2015, p. 886). Additionally, internet connection issues and system operator failure can cause business disruptions. Besides security considerations, regulations might also limit the use of CC as Interview partner 2 highlights. The collaboration with CC providers such as Amazon is restricted because of data security reasons. Additionally, regulators such as the U.S. Food and Drug Administration (FDA) require data availability at any time.

Finally, cloud-based solutions do not necessarily guarantee the interoperability among functions and companies in the SC as Interview partner 3 explains. Even within a cloud-based system, site-specific processes can be established, which could impede interoperability (Interview partner 3). Generally speaking, CC penetration among the interview partners' companies is low (Interview partners 1, 2, 3, 7, and 9). Besides the above-mentioned security concerns, Interview partners 1 and 4 outline two other reasons that might speak against a major IT transformation. Interview partner 1 highlights that larger companies often conduct several M&A activities per year and consequently have a very fragmented IT landscape that takes years to integrate. Interview partner 4 explains that the introduction of new IT systems is not only expensive but also very challenging and he adds that "Such a SAP introduction, for example, [...] can drive companies into their knees".

1.3.2.4.2 Tension: Big Data analytics and artificial intelligence – Prescriptive versus predictive The influx of inexpensive data generated and distributed in real time by objects provides the basis for SCM to utilize this information, transform it into intelligence, and leverage the intelligence to steer the SC. These processes, that were traditionally done by humans, are nowadays increasingly conducted by machines. AI and its subfield machine learning offer enormous potential to optimize SCs (Kersten et al., 2017, p. 12). The speed and accuracy of insights is increased as AI can automatically integrate data from different sources, classify the data, and surface the smallest patterns (Padmos et al., 2016, p. 9). The systems can either forecast what could happen in the future (predictive analytics) or advise on the outcome of future actions (prescriptive analytics) (Kersten et al., 2017, p. 46). The generated insights can be used in two ways: either as inputs for decision makers to make more informed decisions or to automate decisions and thereby processes.

SC planning, for example, demand and supply forecasting and production planning, are recurring and time-consuming activities that are mainly conducted manually (Alicke et al., 2017). With the help of AI systems, the majority of those activities could be automated delivering a higher quality in real time (Alicke et al., 2017). Systems can analyze a variety of different internal (e.g., sensor-generated data) and external (e.g., social media trends, weather and traffic) data, detect and model complex connections, and based on that predict future events (Alicke et al., 2017; Kersten et al., 2017, p. 46; Padmos et al., 2016, p. 10). A human, however, has the final say and can either choose the predicted event or update the information based on his or her experience. This is referred to as predictive analytics.

A typical use case for predictive analytics is demand forecasting, in which additional to internal data also external data such as weather, sales campaigns, and holidays can be integrated to improve forecast results. The forecasted demand can thereafter be fed into capacity planning, routing planning, and optimization as well as procurement optimization (Schmidt et al., 2015, p. 2; Theisinger et al., 2015, p. 10). The usage of predictive analytics can result in the optimization of machine and transport utilization, flow of goods, product availability, and eventually translate into a competitive advantage (Garrett, 2017, p. 23; Kersten et al., 2017, p. 27 and 66; Padmos et al., 2016, p. 10).

Prescriptive analytics goes one step further and enables the automation of decisions. Systems such as Apache Hadoop, IBM Watson, and SAP Hana enable an agile SC that reacts automatically and in real time to sensor generated data (Zillmann, 2016, p. 19). Examples include real-time route optimization considering traffic jams, delivery frequency, and schedule adherence or the forecast of client dissatisfaction and the initiation of corrective actions based on internal and public client data (Zillmann, 2016, p. 19). Whereas automated decisions in definite areas are already possible today, the fully automated SC is still a vision. In his vision, operators only supervise the activities, incidents, and counter measures in the supply network but do not intervene into the system. This is confirmed by Interview partner 3 who considers the future SCM as the supervision of automated processes. With the realization of this vision, sense, and respond, SCs will be transformed into to predict and act systems (Butner, 2010, p. 27).

However, prescriptive analytics as the basis of autonomous decision-making requires not only interdisciplinary IT and technology models, but also a large, diverse, and high-quality data basis (Ivanov & Sokolov, 2012, p. 978; Interview partner 9). Interview partner 9 sees the lack of diverse data as major obstacle of why such systems are still a vision. It is likely that systems to do not generate satisfactory results yet or that companies still fully mistrust automated systems. Therefore, Interview partner 1 is convinced that AI is mainly applied by companies that have a large data basis such as DHL, automotive companies, law firms, and audit companies in the near future. In addition to the lack of data and high investments into IT infrastructure, the loss of jobs needs to be taken into consideration.

Interview partner 8 criticizes that in the context of Industry 4.0 and the digitalization, only the upsides are discussed but that likely job losses due to automation and AI are often disregarded.

1.3.2.4.3 Tension: Blockchain – Wait and see versus orchestrate
Blockchain offers the potential to counter some of today's most pressing SC challenges such as insufficient transparency and therefore low accountability, a lack of interoperability, a high amount of manual paper-based work, and security concerns (Lierow, Herzog, & Oest, 2016; Roeck, Sternberg, & Hofmann, 2019).

Companies can establish end-to-end transparency and accountability in their SCs by "using a blockchain to transfer title and record permissions and activity logs so as to track the flow of goods and services between businesses and across borders" (Casey & Wong, 2017, p. 3). Organizations that depend on a multitude of suppliers to deliver parts can monitor their partners in real time and act if necessary (Casey & Wong, 2017, p. 4). The scalability of the decentralized cloud database facilitates the integration of additional players without impairing data consistency (Lierow et al., 2016). The seamless integration is especially helpful given the ever-increasing complexity of supply networks and the demand for flexible re-configurations.

In light of the need to align disparate interests among SC players, blockchain technology offers another advantage for the usage in SCs: due to the decentralized database, no company is in overall control of the data thereby resolving disclosure and accountability problems among the institutions in the SC (Casey & Wong, 2017, p. 3).

A group of global food companies, among others Walmart, Unilever, and Nestlé, for example, is collaborating with IBM to leverage blockchain in their food SCs (Hackett, 2017a). Given the complex global supply networks consisting of a multitude of diverse players such as farmers, distributors, processors, retailers, and consumers, it can take weeks to spot issues (Hackett, 2017a). Walmart has run a pilot using a Hyperledger Fabric to trace mangoes from Mexico (Hackett, 2017a). On the farm, numerical tags were assigned to mango pallets and their status was recorded in a blockchain at predefined checkpoints throughout the SC up to the store (Hackett, 2017b). Thereby, the time to find out about the source of the mangoes was reduced from nearly a day to about 2 seconds (Hackett, 2017b). A second use case for blockchain technology is smart contracts. Those contracts enable an automatic execution of "contractual rights and obligations, including the terms for payment and delivery of goods and services [...] by an autonomous system that's trusted by all signatories" (Casey & Wong, 2017, p. 4). Interview partners 6 and 8 see great potential to automate negotiation and supplier contracts using smart contracts in combination with blockchain technology.

Finally, users can take advantage of the technology's security. The decentralized storage feature makes cyberattacks particularly difficult. Once a member in the SC records a transaction in a block, the data is nearly non-changeable as it is linked to every previous record (Casey & Wong, 2017, p. 3). Every member of the supply network has a copy of all transactions and can assign specific access rights to ensure confidentiality of private corporate data (Lierow et al., 2016). The secure and provable blockchain technology can be leveraged for a specific movement within Industry 4.0 to protect intellectual property (IP): additive manufacturing. As Interview partner 6 highlights, blockchain technology can prove helpful to exchange data such as "constructional drawing, parameter setting, which are used, for example, to set up a machine, to produce a specific component". By using a blockchain, the owner of the design files can secure the sensible IP especially if more than one party is involved in the process.

However, besides the many advocates that orchestrate the blockchain technology, voices are raised in the research community and among practitioners to rather adopt a wait-and-see attitude. Blockchain is a radical solution and the interviews revealed that the technology is still scarcely discussed in SC functions (Interview partners 2, 4, 5, 6, 8, and 9). Implementing blockchain technologies not only requires adequate specialists, but also the willingness of other SC members. This could prove challenging given the current caveats of the technology.

One obstacle lies in the development of globally accepted technological standards, governance structures, and best practices (Casey & Wong, 2017, p. 5). Additionally, interoperability between private blockchains and public blockchains needs to be established. A big obstacle for agreements across international jurisdictions is the complexity of nationally diverse laws, rules, regulations, and commercial codes (Casey & Wong, 2017, p. 5). Lacking legal security is a major obstacle for companies to operate on blockchain technology. However, even before technological and legal obstacles are approached in a globally coordinated manner, SC functions must define sufficient and especially specific use cases for the technology (Interview partners 4, 7, and 8). Interview partner 4 highlights that he does not see the incremental value of blockchain for logistics, "as data in supply chain is often unclean and needs to be adjusted". Subsequent adjustment of data records in the blockchain is nearly impossible. Additionally, he argues that most SC data is too trivial to justify the security mechanisms of blockchain technology.

However, the interview partners also suggest how the blockchain technology can be deployed advantageously in the SC. Interview partner 9 argues that SCs with very expensive products, such as diamonds or latest drugs, could justify the costs of implementation. The protection of IP and data is outlined by Interview partners 2 and 6. Whereas Interview partner 2 advocates the usage of blockchain for patient cards, Interview partner 6 suggest a blockchain to transfer design files between a developer and an additive manufacturing service contractor.

1.3.2.5 *Technology tensions*

1.3.2.5.1 Tension: SC features – Identify versus instrument Agility has become an important tenet among SC managers (Brusset, 2016, p. 46). Increased competition, SC risks and interruptions, as well as demand fluctuations require flexible, anticipative, and proactive management and execution. Visibility on and the supervision of SC processes are important prerequisites for an agile SCM (Brusset, 2016, p. 56). The SC can be equipped with different technologies that generate diverse types of information to increase visibility. Within the Industry 4.0, those technologies have been integrated into one coherent concept. It is, however, important to highlight that many of those technologies are not Industry 4.0 specific innovations but have rather been introduced several years or even decades ago.

Industry 4.0 and decentral self-steering can only be implemented if objects are distinguishable and uniquely identifiable (Redelberger, 2014). Identification technologies enable the automatic recording of unit movements and status changes, and thereby link material flow and information flow (Raab & Griffin-Cryan, 2011, p. 7). Thus, automatically identifying all units including the products is an important foundation of the digitalized SC (Kersten et al., 2017, p. 26; Redelberger, 2014; Wu et al., 2016, p. 404). Different technologies exist to identify objects with, for example, barcode and radio frequency identification (RFID).

During the literature review and the interviews, it became clear that barcode is especially more prevalent in high flow-rate industries (e.g., Kersten et al., 2017, p. 26; Interview partners 3, 5, 7, 9, and 10; Wu et al., 2016, p. 404). Even though RFID has, prima facie, some advantages such as automatic reading even without direct visual contact, technological issues prevent a widespread application (Kersten et al., 2017, p. 27). Interview partner 3 summarizes the issue by highlighting that "With barcode you could reach 99.8% positive scans and with RFID-Chips only 96% and those three percentage points are sufficient, that the RFID-Chip is disadvantageous in a highly dynamic warehouse with up to 2000 picks per hour". Possible interferences add to the technological issues so that RFID is useful only in delimited areas such as loading device tracking or process bottleneck identification (Kersten et al., 2017, p. 27). Regulatory requirements also limit the use of RFID as Interview partner 2 outlines. Medical agencies such as FDA have to approve the usage of chips on implants. However, he and Interview partner 10 highlight the importance of the application of identification technologies especially for security relevant parts. In sum, barcodes and quick response codes (QR-codes), an advancement that can carry more information, can be applied to increase traceability, transparency, and control manufacturing processes and product quality (Bernhard, 2016).

Besides identifying objects at predefined meter points in the SC, companies can invest into more advanced technological solutions to instrument the SC and increase information transparency beyond identification. Sensor, localization, and mobile communication technologies increasingly generate information that was previously created by humans in the SC (Butner, 2010, p. 26). Applying sensor technologies in warehouses, means of transportation, and containers helps to collect context information such as temperature, filling status, inclination angles, product quality, and humidity (Kersten et al., 2017, p. 27). Localization technologies can be utilized to increase in-transit visibility through real-time tracking of persons, means of transportation, and pallets (Zillmann, 2016, p. 16). Real-time tracking can be used, for instance, for dynamic route management (Zillmann, 2016, p. 16). Real-time access and exchange of those information is enabled by mobile communication technologies that connect objects in the IoT. Interview partner 8 highlights for their digitalized SC that "the first layer is the topic connectivity. That means we have to connect everything".

Interview partner 3 describes a pilot of a Bavarian OEM, its supplier, and logistics providers in which means of transportation and product carriers were equipped with sensors for acceleration, inclination, and temperatures. The OEM gathered the sensor data real time and analyzed the information to deduce actions at goods receipt (Interview partner 3). The same interview partner also highlights, however, that in his opinion a joint effort of the network is necessary to realize such pilots and projects.

1.3.2.5.2 Variable material flows – Supportive technology versus robotic

The idea of the self-steering SC that was touched upon in the previous chapter not only requires a linkage of material and information flows, but also an automation of material flows. Besides the vision of a self-steering SC, cost pressure urges companies to make processes more efficient and unlock saving potentials. Whereas companies started to automate their (standardized) production processes in the context of the third industrial revolution during the 1970s (Kagermann, 2015, p. 32), non-standardized processes are still mainly done manually. SC managers can either semi-automate these non-standardized processes by equipping employees with supportive technologies or completely automate processes with robotic solutions.

One supportive technology is augmented reality, which could be used to semiautomate processes and link information and product flow. The analysis of literature and interviews revealed that supportive technologies could be especially implemented in two cases: if costs are too high for automation or if manual work is appreciated. On the one hand, the high costs of automating complex warehousing and production processes apparently discourage organizations to invest (Interview partners 1, 3, 5, 6, and 7). Interview partner 1 explains that robots "are only relevant and economically viable, if companies have very large facilities, in which a great deal of products are manufactured or handled. An example, I could think of, is Amazon". Other than that, wearables and augmented reality solutions can be applied for expensive and labor-intensive processes that are often paper-based and susceptive to human mistakes such as commissioning (Alicke et al., 2017; Howells, 2016). Interview partner 7 outlines that their warehouse employees are equipped with headsets to receive instructions. On the other hand, automation might not be desirable for specific products such as certain watches and other high-priced lines as Interview partner 9 outlines. Clients expected that "a very expensive watch is indeed manufactured by a skilled watchmaker and that the manufacturing process is automated as little as possible".

If scale and product features, however, allow an automation of material flows, the investment risk can be minimized through cooperation and partnerships. Interview partner 4 explains that a pilot was started in the organization he is working for in cooperation with a German start-up that develops and manufactures commissioning robots. The robots "learn" the commissioning of similar sized squared boxes in one of their warehouses. In his opinion, the introduction of robots is inevitable given the increasing shortage of warehouse logistics employees. Other use cases include labeling, palletizing, as well as loading and unloading (Zillmann, 2016, p. 15).

Investment costs for a fully automated warehouse or forklift trucks for differently sized and shaped products are nonetheless very high. Reduced operating cost, lead times, and environmental impact as well as increased accuracy and efficiency could however make the investment worthwhile in the long term (Alicke et al., 2017; Interview partners 3 and 8). A critical part will be the design of the so-called man and machine interaction (Kersten et al., 2017, p. 21) as well as the handling of rationalized jobs (Interview partner 8).

Besides automated manufacturing and intra-logistics, external delivery is increasingly considered to be automated. Whereas the introduction of fully autonomously driving trucks is still facing technological and regulatory challenges, automated solutions for last mile delivery are increasingly tested. During the interviews it became clear, however, that the delivery via small robots or drones might remain a niche application (Interview partners 2 and 4).

1.3.2.5.3 Additive manufacturing – Niche versus integrated usage Originally used for rapid prototyping, additive manufacturing (AM) nowadays has the potential to transform many aspects of the SC including its configuration, players and processes (Ben-Ner & Siemsen, 2017, p. 5; Stölzle et al., 2017, p. 98). Companies along the SC are increasingly exploring the implications of an AM adoption. Besides leveraging the technology for small-scale or niche productions (e.g., spare parts or prototypes), an integrated industrial-scale usage of AM technologies is increasingly discussed (Oettmeier & Hofmann, 2017). Proponents of the technology argue that its implementation helps to cope with some of today's most prevalent SC challenges such as demand fluctuations, need for individualization, and mass customization and sustainability (Ben-Ner & Siemsen, 2017, p. 11).

AM technologies can be added to conventional production systems for niche applications such as prototypes or spare parts. Interview partner 9 sees use cases for spare parts for products reaching the end of the lifecycle for which a reactivation of a production line or a major warehouse would be uneconomic. Thereby, manufacturers could decrease their spare part inventory and the related costs (Schrauf & Berttram, 2016, p. 24). Moreover, prototyping the construction of spare parts of non-standard parts (Interview partner 3) and the printing of "small series, which need to be very specific, for which the requirements need to be adapted often, for which a final product description is lacking" (Interview partner 9) are predestined for AM technologies.

Associated implementation costs, the range of printable raw materials, and regulatory aspects are major factors that suggest a niche application for AM technologies (Schmidt et al., 2015, p. 11; Interview partners 2 and 10). A survey among SC managers shows that the limited selection of printable raw materials is constraining the replacement of traditional production systems by AM technologies (Schmidt et al., 2015, p. 11). Regulatory aspects narrow the application scope in certain industries such as medical devices and public transportation (Interview partners 2 and 10).

Besides niche applications, organizations along the SC can make AM technologies an integral part of their value creation process. For the detailed description of use cases, it is essential to differentiate between manufacturing companies and those that have not constructed products hitherto. As shown above, the construction of low-volume spare parts is a first step into AM technologies for manufacturing companies. The technology can also be used for high-volume spare part production as Interview partner 10 highlights, "We have a 3D printing project ongoing, especially for non-safety critical parts, to react flexibly to demand changes". The interview partner adds that also safety-critical parts could be constructed with AM technologies as soon as stability, durability, and period of application are better understood. The printers can also be placed at the client's site, if the demand for spare parts is high and critical, to rapidly construct replacements and thereby decrease downtime of equipment (Mussomeli et al., 2016, p. 6; Schrauf & Berttram, 2016, p. 23).

However, not only spare parts can be produced at the client's site, but also final products as Interview partner 2 outlines when saying that "it is not unlikely, that parts are printed directly in the hospital or at any other point, that is closer to the client, and there have already been custom-made items, that are based on that technology".

Other factors to consider are packaging costs and environmental impacts. The decentralized production close to the client can help to reduce the ecological footprint as products and spare parts, especially in low volumes, are not necessarily shipped around the globe (Interview partner 9). Interview partner 8 points to leveraging AM technologies for packaging to decrease this major pool of costs.

This "make-to-use" (Mussomeli et al., 2016, p. 11) with AM technologies could have far-reaching implications on the SC which should not be underestimated: locations and materials change, SCs shorten, and configurations become more flexible (Schmidt et al., 2015, p. 2). Consequently, the business model of logistics companies is, at least partially, endangered as the digital delivery of production data would cut the physical delivery of products (Thomas et al., 2016, p. 60). Nevertheless, logistics companies could leverage AM technologies by adapting their business model and equipping, for instance, warehouses with AM technologies (Thomas et al., 2016, p. 62; Zillmann, 2016, p. 29). By taking production activities into the product portfolio, wholesalers could also benefit from AM technologies and increase their share in the value adding process within the SC (Thomas et al., 2016, p. 61).

AM technologies, however, are not suitable for all product types. Interview partner 6 outlines that AM technologies are likely to become accepted for products too complex to be produced by conventional manufacturing technologies or products for which a small scale is economical. Given the constant technological progress of the AM technologies, the economic threshold is constantly lowered. In order to test the applicability of the technology, interested companies could draw on the experience of AM providers, collaborate on small pilot productions, and decide based on the manufacturing and investment costs if the introduction is economically viable.

The 14 tensions described above show that decisions makers in SCM functions are confronted with a diverse and challenging set of tasks in light of the SC digitalization. The tensions are cross-functional and cover strategy and business model, operating model and execution, organization, data and IT, as well as new technologies. The in-depth analysis shows that each tension holds the potential to be either an opportunity or a challenge. Decision makers and SCM employees need to analyze each tension in light of the company's characteristics, such as business model, industry, and position in the SC, to comprehend the full implications of the digitalization on SCM.

1.4 Implications for practice and research

1.4.1 Implications for practice

Schrauf and Berttram (2016, p. 4) suggest that if the vision of Industry 4.0 can be realized, SCs become more cost-efficient, customer-oriented, and agile while new business opportunities could open up. Nevertheless, before potentials can be unlocked, the empirical results indicate that companies are likely to encounter diverse challenges and obstacles on their digitization path. The chapter on hand has pinpointed and categorized the most important digital SCM challenges and presented potential solutions. The results of the empirical analysis suggest that the implications for managers are threefold. In a first step, companies need to analyze their starting position by determining their current digitalization degree of SCM activities. The analysis of the status quo is essential for identifying the gaps toward the target status of a digitalized SCM. The second step entails defining the target status by developing a digital SCM vision (Mussomeli et al., 2016, p. 9; Schrauf & Berttram, 2016, p. 30; Interview partners 3 and 8). The tensions introduced through the tension framework can be used as guiding principles for structuring and developing a digital SCM vision. The framework, however, should only serve as a starting point and be augmented by other inputs such as traditional strategy research. In particular, the company's position in the SC, its company strategy, and its industry characteristics need to be taken into consideration. The starting position and the target status help to identify the biggest and most pressing gaps. In a third step, appropriate solutions to close the gaps and to achieve the target status need to be developed. The tension framework and the detailed tension descriptions introduce possible solutions to close those gaps. The empirical analysis indicates that decision makers need to develop an appropriate roadmap by prioritizing activities and identifying quick wins (Schrauf & Berttram, 2016, p. 30). Additionally, the development and deployment of small pilots in zoned business environments before rolling out company-wide projects seem to be most promising (Schrauf & Berttram, 2016, p. 30; Interview partners 3, 4, and 8).

Ultimately, the setup of a two-tier architecture, including a dedicated digital (SCM) business unit to give the digitalization effort sufficient importance within the organization and to pool initiatives is promising (Alicke et al., 2017; Interview partners 1, 3, and 8). Such a business unit should be equipped with sufficient freedom to experiment and cash to invest (Interview partners 1 and 8). Ideas developed within this business unit can be piloted in the organization and afterward rolled out company-wide.

In sum, it becomes clear that tackling digital SCM challenges is not a one-off exercise but rather a transformation process that needs to be planned in the long-term and henceforth step-wise implemented.

1.4.2 Implications for research and limitations

The chapter on hand contributes to the SCM research with an empirical methodology based on a structured literature review and semi-structured interviews. Moving forward, some limitations are acknowledged and an outlook on future research is given. First, during the selection phase for the structured literature review most likely not all relevant papers and articles have been identified. This might be due to incompleteness of the predefined key terms or the usage of alternative terms in relevant articles. Additionally, scientific research at the interface of SCM and digitalization is still thin. Second, the sample size of ten for a structured literature review gives a good overview of the topic. However, increasing the sample size and choosing an alternative sample method given a longer research time frame could have strengthened the external validity of the results. Third, this chapter is focused on identifying challenges and solutions for digitalizing SCM in general but is lacking a specific focus on industries, companies, and positions in the SC. Another way forward would be the analysis within a specific industry or position in the SC or with a focus on company strategies. The usage of case studies could be one way to distill insightful findings. Moreover, a quantification in terms of urgency and relevance is missing given the limited research time frame. Supplementing the qualitative tension framework with quantitative analyses could reveal insightful findings on how to prioritize and address the tensions. Lastly, the tension framework gives a structured overview of the most important digital SCM challenges. The objective of this chapter is to give a broad overview of the current digital SCM challenges and thus said challenges and potential solutions are only analyzed and described in general. Therefore, this chapter opens up several new and interesting research streams, particularly focusing on each individual tension elaborated in this work. An in-depth analysis of single tensions, such as the implementation of transparency in the SC or the introduction of AM technologies could be though-provoking and help to develop detailed solutions. While some promising publications for the tensions exists, at-scale scientific research is still scarce.

1.5 Conclusion

With the Hannover Messe in 2011, the promising impact of digital technologies on operations management in general and SCM in particular have come to the fore. Ever since, a lot of attention in the mainstream business press and scientific literature has been given to the vast potentials, the digitalization of SCM activities shows promise. Less focus, however, has been dedicated to challenges and obstacles that decision makers might encounter on their digitalization path. We believe, this chapter is first to execute an in-depth and comprehensive analysis of the most important challenges and solutions for digitalizing SCM. By answering the main research question, the chapter contributes to the current lively discussion of digital SCM in three ways.

First, the results of the chapter highlight the 14 most important digital tensions. Second, each tension is described in-depth, thereby equipping the reader with a more detailed understanding of current challenges and obstacles. Moreover, an option space of potential solutions for each tension is developed based on literature insights and real-life examples of the interview partners. Third, the empirical findings are reflected in light of the current SCM research and potential impacts of the ongoing digitalization on the current understanding and characteristics of SCM are highlighted.

In summary, the digitalization of SCM holds vast potentials for organizations and customers around the globe. However, in order to tap the full potential, decision makers need to address various challenges and obstacles. This chapter acts as an excellent starting point for identifying the most important challenges and potential solutions to digitalize SCM.

References

Alicke, K., Rexhausen D., & Seyfert, A. (2017). Supply Chain 4.0 in Consumer Goods. Retrieved from http://www.mckinsey.com/industries/consumer-packaged-goods/our-insights/supply-chain-4-0-in-consumer-goods

Bengtsson, M. (2016). How to plan and perform a qualitative study using content analysis. NursingPlus Open, 2, 8–14. doi:10.1016/j.npls.2016.01.001

Ben-Ner, A., & Siemsen, E. (2017). Decentralization and localization of production: The organizational and economic consequences of additive manufacturing (3D printing). California Management Review, 59(2), 5–23. doi:10.1177/0008125617695284

Benton, D. (2017). Supply Chain 4.0: Adidas and Amazon Re-Write the Rules on Supply Chain Management. Retrieved from http://www.supplychaindigital.com/scm/supply-chain-40-adidas-and-amazon-re-write-rules-supply-chain-management

Bernhard, R. (2016). Industrie 4.0 – das Supply Chain Management der Zukunft. Retrieved from https://www.bigdata-insider.de/industrie-40-das-supply-chain-management-der-zukunft-a-524098/

Bettis, R. A., Gambardella, A., Helfat, C., & Mitchell, W. (2015). Qualitative empirical research in strategic management. Strategic Management Journal, 36(5), 637–639. doi:10.1002/smj.2317

Brusset, X. (2016). Does supply chain visibility enhance agility? International Journal of Production Economics, 171(1), 46–59. doi:10.1016/j.ijpe.2015.10.005

Butner, K. (2010). The smarter supply chain of the future. Strategy & Leadership, 38(1), 22–31. doi:10.1108/10878571011009859

Casey, M. J., & Wong, P. (2017). Global supply chains are about to get better, thanks to blockchain. Harvard Business Review, 2–6. Retrieved from https://hbr.org/2017/03/global-supply-chains-are-about-to-get-better-thanks-to-blockchain

Cooper, M. C., Ellram, L. M., Gardner, J. T., & Hanks, A. M. (1997). Meshing multiple alliances. Journal of Business Logistics, 18(1), 67–89.
Cooper, M. C., Lambert, D. M., & Pagh, J. D. (1997). Supply chain management: More than a new name for logistics. The International Journal of Logistics Management, 8(1), 1–14. doi:10.1108/09574099710805556
Day, G. S. (2011). Closing the marketing capabilities gap. Journal of Marketing, 75(4), 183–195. doi:10.1509/jmkg.75.4.183
Downe-Wamboldt, B. (1992). Content analysis: Method, applications, and issues. Health Care for Women International, 13(3), 313–321. doi:10.1080/07399339209516006
Elo, S., & Kyngas, H. (2008). The qualitative content analysis process. Journal of Advanced Nursing, 62(1), 107–115. doi:10.1111/j.1365-2648.2007.04569.x
Eßig, M., Hofmann, E., & Stölzle, W. (2013). Supply Chain Management. München: Vahlen.
Fayezi, S., Zutshi, A., & O'Loughlin, A. (2016). Understanding and development of supply chain agility and flexibility: A structured literature review. International Journal of Management Reviews, 19(4), 379–407. doi:10.1111/ijmr.12096
Fischer, M. (2017). Auf Dem Weg Zur Digitalisierten Industrie. Retrieved from https://www.gs1network.ch/schwerpunkt/2017/supply-chain-4-0,-1-2017/item/1592-auf-dem-weg-zur-digitalisierten-industrie.html
Garrett, R. (2017). The benefits of digital transformation. Supply & Demand Chain Executive, 18(2), 22–25.
Gstettner, S., Roesgen, R., Ganeriwalla, A., Kotlik, L., & Walter, G. (2016). Three Paths to Advantage with Digital Supply Chains. Retrieved from https://www.bcgperspectives.com/content/articles/supply-chain-management-technology-digital-three-paths-advantage-digital-supply-chains/
Hackett, R. (2017a). Walmart and 9 Food Giants Team Up on IBM Blockchain Plans. Retrieved from http://fortune.com/2017/08/22/walmart-blockchain-ibm-food-nestle-unilever-tyson-dole/
Hackett, R. (2017b). Why Big Business is Racing to Build Blockchains. Retrieved from http://fortune.com/2017/08/22/bitcoin-ethereum-blockchain-cryptocurrency/
Handfield, R. (2016). Preparing for the era of the digitally transparent supply chain: A call to research in a new kind of journal. Logistics, 1(2).
Hanifan, G., Crosnier, S., & Timmermans, K. (2016). Digital Trendsetters: Secrets of the Most Successful Digital Supply Chains. Retrieved from https://www.accenture.com/t20161216T015905Z__w__/us-en/_acnmedia/PDF-26/Accenture-Strategy-Digital-Trendsetters-Successful-Supply-Chain.pdf#zoom=50
Harwood, T. G., & Garry, T. (2003). An overview of content analysis. Marketing Review, 3(4), 479–498. doi:10.1362/146934703771910080
Hofmann, E. (2007). The flow of financial resources: An inevitable part of supply chain design activities. In W. Delfmann & T. Klaas-Wissing (Eds.), Strategic Supply Chain Design (pp. 173–201). Köln: Kölner Wissenschaftsverlag.
Hofmann, E., & Rüsch, M. (2017). Industry 4.0 and the current status as well as future prospects on logistics, Computers in Industry, 89, 23–34. https://doi.org/10.1016/j.compind.2017.04.002
Howells, R. (2016). The Supply Chain Goes Digital: Opportunities and Challenges. Retrieved from https://blogs.sap.com/2016/05/18/the-supply-chain-goes-digital-opportunities-and-challenges/

Hu, M., & Monahan, S. T. (2015). Sharing supply chain data in the digital era. MIT Sloan Management Review, 57(1), 95–96.
International Air Transport Association. (2017). Multilateral e-AWB Agreement. Retrieved from http://www.iata.org/whatwedo/cargo/e/eawb/Pages/multilateral.aspx
Ivanov, D., & Sokolov, B. (2012). The inter-disciplinary modelling of supply chains in the context of collaborative multi-structural cyber-physical networks. Journal of Manufacturing Technology Management, 23(8), 976–997. doi:10.1108/17410381211276835
Jacques, A. (2017). The digital supply chain: Seizing pharma's untapped opportunity. Pharmaceutical Technology, 2017(1), 20–23.
Jede, A., & Teuteberg, F. (2015). Integrating cloud computing in supply chain processes: A comprehensive literature review. Journal of Enterprise Information Management, 28(6), 872–904. doi:10.1108/JEIM-08-2014-0085
Jick, T. D. (1979). Mixing qualitative and quantitative methods: Triangulation in action. Administrative Science Quarterly, 24(4), 602–611. doi:10.2307/2392366
Jocher, G. (2017). Logistik 4.0: Digitalisierung der Wertschöpfungskette. Retrieved from https://www.sazsport.de/markt-sortimente/sazsport-experten/logistik-4.0-digitalisierung-wertschoepfungskette-1235236.html
Kache, F., & Seuring, S. (2017). Challenges and opportunities of digital information at the intersection of big data analytics and supply chain management. International Journal of Operations & Production Management, 37(1), 10–36. doi:10.1108/IJOPM-02-2015-0078
Kagermann, H. (2015). Change through digitization: Value creation in the age of industry 4.0. In A. Pinkwart, R. Reichwald, H. Albach, & H. Meffert (Eds.), Management of Permanent Change (pp. 23–45). Wiesbaden: Gabler.
Kagermann, H., Lukas, W.-D., & Wahlster, W. (2011). Industrie 4.0: Mit dem Internet der Dinge auf dem Weg zur 4. industriellen Revolution, Retrieved from http://www.wolfgang-wahlster.de/wordpress/wp-content/uploads/Industrie_4_0_Mit_dem_Internet_der_Dinge_auf_dem_Weg_zur_vierten_industriellen_Revolution_2.pdf
Kersten, W., Seiter, M., See, B. v., Hackius, N., & Maurer, T. (2017). Trends und Strategien in Logistik und Supply Chain Management – Chancen Der Digitalen Transformation. Retrieved from https://logistiktrends.bvl.de/system/files/t16/2017/Trends_und_Strategien_in_Logistik_und_Supply_Chain_Management_-_Chancen_der_digitalen_Transformation_-_Kersten_von_See_Hackius_Maurer_2017.pdf
Kohlbacher, F. (2006). The use of qualitative content analysis in case study research. Forum: Qualitative Social Research, 7(1), 1–30. doi:10.17169/fqs-7.1.75
Laursen, M., & Svejvig, P. (2016). Taking stock of project value creation: A structured literature review with future directions for research and practice. International Journal of Project Management, 34(4), 736–747. doi:10.1016/j.ijproman.2015.06.007
Leeflang, P. S. H., Verhoef, P. C., Dahlström, P., & Freundt, T. (2014). Challenges and solutions for marketing in a digital era. European Management Journal, 32(1), 1–12. doi:10.1016/j.emj.2013.12.001
Lierow, M., Herzog, C., & Oest, P. (2016). Blockchain: The Backbone of Digital Supply Chains. Retrieved from http://www.oliverwyman.com/our-expertise/insights/2017/jun/blockchain-the-backbone-of-digital-supply-chains.html

Madhani, P. M. (2017). Supply chain strategy selection: A multi-criteria decision-making approach. IUP Journal of Supply Chain Management, 14(2), 38–56.

Majeed, M. A. A., & Rupasinghe, T. D. (2017). Internet of things (IoT) embedded future supply chains for industry 4.0: An assessment from an ERP-based fashion apparel and footwear industry. International Journal of Supply Chain Management, 6(1), 25–40.

Mayring, P. (2014). Qualitative Content Analysis: Theoretical Foundation, Basic Procedures and Software Solution. Retrieved from https://www.ssoar.info/ssoar/bitstream/handle/document/39517/ssoar-2014-mayring-Qualitative_content_analysis_theoretical_foundation.pdf?sequence=1

Mussomeli, A., Gish, D., & Laaper, S. (2016). The Rise of the Digital Supply Network – Industry 4.0 Enables the Digital Transformation of Supply Chains2017. Retrieved from https://dupress.deloitte.com/dup-us-en/focus/industry-4-0/digital-transformation-in-supply-chain.html

Oettmeier, K., & Hofmann, E. (2017). Additive manufacturing technology adoption: an empirical analysis of general and supply chain-related determinants. Journal of Business Economics, 87, 97–124. https://doi.org/10.1007/s11573-016-0806-8

Padmos, D., Alexander, M., Brody, P., Chadam, J., Cookson, C., Little, J., & Meadows, B. (2016). Digital Supply Chain: It's All about That Data. Retrieved from http://www.ey.com/Publication/vwLUAssets/Digital_supply_chain_-_its_all_about_the_data/$FILE/EY-digital-supply-chain-its-all-about-that-data-final.pdf

Papert, M., & Pflaum, A. (2017). Development of an ecosystem model for the realization of internet of things (IoT) services in supply chain management. Electronic Markets, 27(2), 175–189. doi:10.1007/s12525-017-0251-8

Pfeiffer, S. (2017). The vision of "industrie 4.0" in the making – a case of future told, tamed, and traded. NanoEthics, 11(1), 107–121. doi:10.1007/s11569-016-0280-3

Raab, M., & Griffin-Cryan, B. (2011). Digital Transformation of Supply Chains: Creating Value - When Digital Meets Physical. Retrieved from https://www.capgemini.com/wp-content/uploads/2017/07/Digital_Transformation_of_Supply_Chains.pdf

Redelberger, J. (2014). Industrie 4.0 im Kontext Logistik – Connected Supply Chains. Retrieved from https://www.capgemini.com/consulting-de/2014/08/industrie-40-im-kontext-logistik-connected-supply-chains/

Roeck, D., Sternberg, H., & Hofmann, E. (2019). Distributed ledger technology in supply chains: A transaction cost perspective, International Journal of Production Research. doi: 10.1080/00207543.2019.1657247

Rowe, F. (2014). What literature review is not: Diversity, boundaries and recommendations. European Journal of Information Systems, 23(3), 241–255. doi:10.1057/ejis.2014.7

Schmidt, B., Rutkowsky, S., Petersen, I., Klötzke, F., Wallenburg, C. M., & Einmahl, L. (2015). Digital Supply Chains: Increasingly Critical for Competitive Edge. Retrieved from https://www.atkearney.com/documents/20152/435077/Digital%2BSupply%2BChains.pdf/82bf637e-bfa9-5922-ce03-866b7b17a492

Schrauf, S., & Berttram, P. (2016). How Digitization Makes the Supply Chain More Efficient, Agile, and Customer-Focused. Retrieved from https://www.strategyand.pwc.com/media/file/Industry4.0.pdf

Schurk, H.-E., Harth, S., Herrmann, L., Rosina, P., & Scheid, C. (2015). Cloud-Computing als Chance für Unternehmen. Potentiale der Cloud und wie sie von KMU genutzt werden können – Leitfaden. Retrieved from http://search.ebscohost.com/login.aspx?direct=true&db=edswis&AN=edswis.BEFO20160309066.M.WTI..BEFO.DOMA.ZDEE⟨=de&site=eds-live&authtype=ip,uid

Stölzle, W., Hofmann, E., & Mathauer, M. (2018). Logistikmarktstudie Schweiz Bands 2018: Autonom und vernetzt—Supply Chain-Entwicklungen im digitalen Zeitalter. Bern: GS1 Schweiz.

Stölzle, W., Hofmann, E., & Oettmeier, K. (2017). Logistikmarktstudie Schweiz Band 2017: Logistik und Supply Chain Management im Zeitalter der Digitalisierung. Bern: GS1 Schweiz

The Economist Intelligence Unit Limited. (2017). Rebooting Supply Chains: Shorter, Smarter and More Sustainable? Retrieved from http://growthcrossings.economist.com/wp-content/uploads/sites/47/2017/05/eiu-growth-crossings-treasury-pillar-briefing-paper-final-may-2017.pdf

Theisinger, F., Specht, J., & Hasenstab, H. (2015). Agilität Durch Digitalisierung: Supply Chain Management in der digitalen Wirtschaft. Retrieved from https://www.detecon.com/sites/default/files/BME_Kongress_DeteconBroschüre_2015.pdf

Thomas, O., Kammler, F., Zobel, B., Sossna, D., & Zarvic, N. (2016). Supply chain 4.0: Revolution in der logistik durch 3D-druck. IM+io Fachzeitschrift für Innovation, Organisation und Management, 31(1), 58–63.

Tranfield, D., Denyer, D., & Smart, P. (2003). Towards a methodology for developing evidence-informed management knowledge by means of systematic review. British Journal of Management, 14(3), 207–222. doi:10.1111/1467-8551.00375

Vendrell-Herrero, F., Bustinza, O. F., Parry, G., & Georgantzis, N. (2017). Servitization, digitization and supply chain interdependency. Industrial Marketing Management, 60, 69–81. doi:10.1016/j.indmarman.2016.06.013

Webster, J., & Watson, R. T. (2002). Analyzing the past to prepare for the future: Writing a literature review. MIS Quarterly, 26(2), xiii–xxiii. doi:10.2307/4132319

Wu, L., Yue, X., Jin, A., & Yen, D. C. (2016). Smart supply chain management: A review and implications for future research. The International Journal of Logistics Management, 27(2), 395–417. doi:10.1108/IJLM-02-2014-0035

Zillmann, M. (2016). Keine Industrie 4.0 ohne Digitalisierung der Supply Chain – Intelligente Logistikdienstleistungen für die Fertigungsindustrie. Retrieved from http://luenendonk-shop.de/out/pictures/0/lue_whitepaper_lhis_sc_f300816_fl.pdf

chapter two

Industry 4.0 creating a buzz in the western hemisphere

But watch out for China pulling into the fast lane

Stefan Heng
Cooperative State University Mannheim, Mannheim, Germany

Contents

2.1 *Industry 4.0: Hype or hope*

All around the globe, Industry 4.0 (also known as integrated industry) is on everyone's lips. In Germany, big companies, the Mittelstand, and certain members of the public are examining the new opportunities associated with it (see Schneider, 2016). This interest has undoubtedly been boosted not only by the focus that Industry 4.0 received by Hanover Fair, the major international industry fair, but also by the government's support worldwide, which besides raising its profile at different summits also includes remarkable grants and public initiatives (see De Propris and Bailey, 2018, pp. 47–48).[1] However, in the vast expanse encompassing augmented reality (see Gushima and Nakajima, 2017; Heng, 2019b, p. 33; or Rauschnabel and Ro, 2016, p. 124), Big Data, cloud computing, cyber-physical systems, RFID chips, Internet of Things and services (see Ezell et al., 2018, p. 17), machine-to-machine communication, Second Machine Age (see Brynjolfsson and McAfee, 2014; p. 10), and Smart X (i.e. intelligent features in any things, e.g. products, grids, factories, mobility, etc.) (see Yudono et al., 2019, p. 330), the term Industry 4.0 remains imprecise (see Kagermann, 2014, p. 604) – which is possibly the intention of many marketing strategists. This imprecise definition of Industry 4.0 repeatedly gives rise to overblown expectations, which lead to disappointment. Providers advertise Industry 4.0 offerings as completing automation and thus enabling cost-effective adaptation of production to individual requirements, as well (see Heng, 2015, p. 11).

The public relations (PR) campaigns in many countries around the globe surrounding Industry 4.0 has ensured that companies and institutions all around the globe cannot get around being aware of the concept for a while (see Staufen, 2019, p. 7; or Papula et al., 2019, p. 105). However, with expectations raised by this campaign what happens all too often is that the fundamental idea of boosting efficiency via sensible automation gets lost in the hype surrounding the buzzword.

Empirical experience shows that companies with a less established process structure (greenfield investments) are more likely to embrace the new elements of Industry 4.0. Modern industrial structures that were built up comparatively recently are well fitted for the ongoing digital evolution. If true, this would have severe implications for countries' competitive advantage on international markets. China sets out to play a leading role in this digital evolution by a comprehensive upgrade of the entire Chinese value chain. The country is determined to seize the outstanding

[1] Public initiatives like "KI Made in Germany", a program arranged by German government in 2018. There, the government will spend EUR 3 billion for AI projects in Germany until 2025. For comparison, in Tianjin, a harbour city in north-east China, alone, they launched a funds ad valorem of EUR 15 billion for AI projects. Thus, five times as much as in entire Germany.

opportunity at hand, as the "Made in China 2025" program underlines, flanked by intensive legislative support for innovative enterprises. The affaire about Meng Wanzhou, CFO of Huawei, who was jailed in Canada in 2018. This gives an impression of the very close relationship between economy and administration.

This study analyses the economic potential of Industry 4.0. The first section examines the principles of Industry 4.0, identifying the most important terms and concepts as well as the key drivers and obstacles. The second section presents business ideas in the Industry 4.0 environment. The third section describes the differences between Germany and China's way towards Industry 4.0. In this regard, the section presents policy actions initiated by the Chinese government. In particular, we look at private-public cooperation to enable and facilitate company-level action, information and communications technology (ICT) infrastructure, workforce quality, and a sound legal framework of data security and data protection (see Ahrens and Spöttl, 2015, pp. 187–190; Albrecht, 2016, pp. 5–8; or Odenbach et al., 2017, pp. 20–22). The final section presents a conclusion and a forecast.

2.2 *Principles of Industry 4.0*

2.2.1 *Next stages of development ahead*

The stages of industrial development have different names depending on the geographical location and the area under examination (e.g. environment, technology) (see Bauernhansl, 2014; or Brynjolfsson and McAfee, 2014). In this analysis, we seek to follow the logic of industrial development, which was developed especially in Germany in close cooperation with the academy of technical sciences, acatech (the National Academy of Science and Engineering), and the Forschungsunion Wirtschaft – Wissenschaft (see Anon., 2013a, p. 23).

According to this, the first industrial revolution commenced at the end of the 18th century with the introduction of mechanical production equipment, such as the mechanical loom for goods manufacturing. With the advent of electrically powered machinery used for mass production based on the division of labour came the second industrial revolution at the turn of the 20th century. The third industrial revolution then commenced in the 1970s. It was based on the use of electronics and information technologies to automate production processes. This entailed machinery performing not only a large proportion of previously manual tasks, but also a number of the intellectual tasks.

The fourth industrial revolution, or Industry 4.0 for short, is to become a reality in the coming decade (see Gornig and Schiersch, 2015, p. 38). The political debate about the term Industry 4.0 focuses equally on the important and abstract objectives. For its promoters, Industry 4.0 is not only about improving Germany's international competitiveness, it is also seen as a tool for tackling the most pressing global challenges (e.g. the consumption of renewable and non-renewable resources) as well as specific national challenges (e.g. the labour supply that is changing due to demographic shifts) (Figure 2.1).

2.2.2 *Source of remarkable fascination*

Industry 4.0 is focused on smart products, procedures, and processes (smart production). A key element of Industry 4.0 is therefore the smart factory. The smart factory controls the fast-growing complexity while also boosting production efficiency. In the smart factory, there is direct communication between man, machine, and resources (see Mertens, 2014, p. 29). Smart products know their manufacturing process and future application. With this knowledge, they actively support the production process and the documentation (when was I made, which parameters am I to be given, where I am supposed to be delivered).

With its interfaces to smart mobility, smart logistics, and smart grids, the smart factory is an important element of future smart infrastructures.

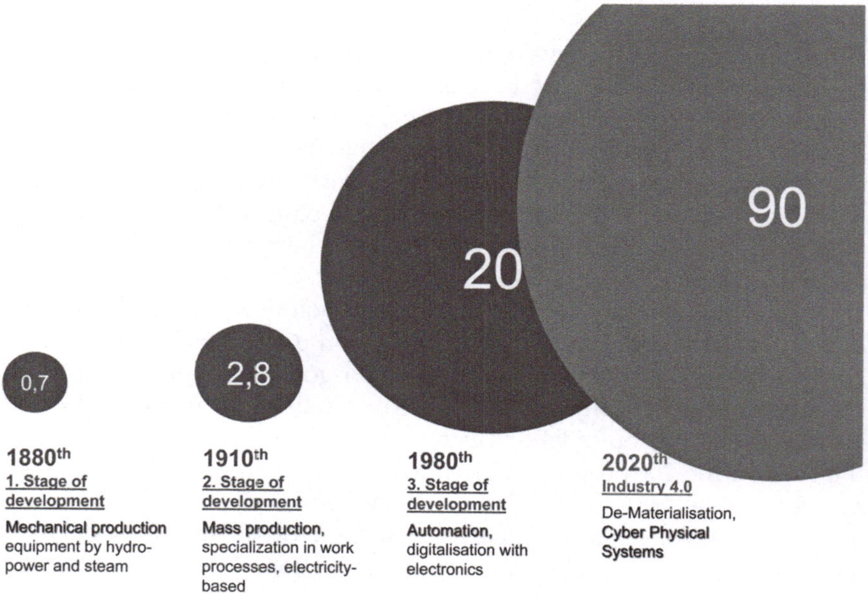

Figure 2.1 Development of economic output concerning economic revolutions: gross domestic product (GDP) worldwide, USD trillion (purchasing power parity adjusted). (Source: Author's illustration, according to Anon., 2013b, pp. 28–29.)

Conventional value chains will thereby be refined and very new business models will become established. Thus, Bitkom (the association for the German ICT industry) and Fraunhofer IOA (public research institute that focuses on investigation of current topics in the field of technology management) predict that the industry could boost their productivity by 30% with the aid of Industry 4.0 (see Bauer et al., 2014, p. 14).

However, essentially, the benefits of Industry 4.0 are not limited to manufacturing. Logistics, financial services, and healthcare are further sectors that will experience rapid change and substantial economic impact (see Ahrens and Spöttl, 2015, p. 188). Often overlooked, agriculture will also be a major beneficiary of the vertical and horizontal integration in the value chain. Moreover, the Internet of Things and services can be used to improve crop irrigation, environmental monitoring, market circulation of agricultural products, and traceability. By doing so, Industry 4.0 will probably have a considerable impact on the economic livelihoods of millions of producers and certainly on billions of customers (Figure 2.2).

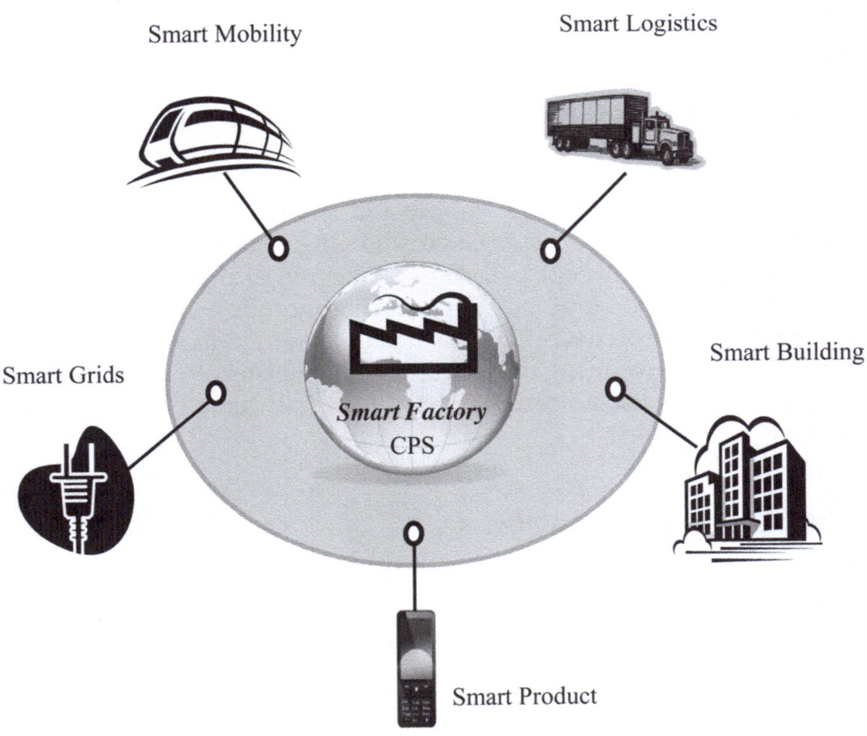

Figure 2.2 In the centre of Industry 4.0. (Source: Author's illustration, according to Anon., 2013b, p. 23.)

2.2.3 Economic boundaries becoming blurred

Basically, Industry 4.0 describes the fourth stage of industrial development, with increasingly smart systems being formed in ever more closely integrated value chains. The Industry 4.0 compliant production facility is thus a completely integrated smart environment (see Heininger, 2018, p. 270).

Accordingly, the implementation of Industry 4.0 seeks to leverage existing technological and market potential, to tap it in a systematic innovation process, and to bring this overall concept together with expertise, services, and the knowledge of employees to create an optimised whole. Companies embarking on the road to Industry 4.0 will pay particular attention to the following aspects:

- Vertical integration of the necessary process stages along the value chain.
- Horizontal integration at one stage of the value chain.
- Seamless end-to-end digital information flow across the entire value chain.

The Industry 4.0 concept must therefore encompass not only value creation per se, but also work organisation, business models, and downstream services. It does this by using information technology to link up production, marketing, and logistics, and thereby captures all resources, production facilities, and warehousing systems. The reorganisation thus extends from the energy supply and smart power grids through to advanced mobility concepts (smart mobility, smart logistics). On the technical side, the concept is based on integrating cyber-physical systems into production and logistics and the rigorous end-to-end implementation of the Internet of Things and services in industrial processes. In this smart environment, the concept of the Internet of Things and services that was devised already a decade ago will actually now become a reality.

2.2.4 Cost-cutting possibilities

The user firms want Industry 4.0 to help them implement cost-efficient manufacturing. What is new here is that the optimisation is performed constantly during ongoing operation and across the entire value chain. The main areas where such potential lies are as follows:

- Capital costs: Companies that optimise their value chains and increase their manufacturing automation thereby reduce their tied-up capital.

- Energy costs: Companies can cut their energy costs via efficient use and smart control of their plant facilities. Many companies pay little attention to this cost item, but it is usually quite a significant sum (Odenbach et al., 2017, p. 30).
- Personnel costs: Companies with highly automated production processes tend to require a declining number of low-skilled employees.

Making a sound assessment of the overall effect (including the costs of training, implementation, and maintenance) within the company will prove difficult in most cases, though. This is particularly the case because numerous user companies, especially small- and medium-sized enterprises (SMEs), can often only very vaguely estimate their actual costs (and thus also the necessity to undertake restructuring).

2.2.5 *More than narrow-minded cost-cutting focus*

Industry 4.0 is, however, by no means limited to the area of costs. Rather, the concept is broad in scope and is described by experts as extremely relevant. For instance, in what is undoubtedly an optimistic projection, acatech asserts that firms could boost their productivity by 30% with the aid of Industry 4.0.

The main parameters that open up the potential for this increase are flexibility, lead times, batch size, and new services and work structuring. We shall address these parameters below (see Heng, 2019a, p. 574):

- More flexibility: Networking enables business processes to be structured more dynamically. Production procedures in particular are to react more flexibly to changes in demand or breakdowns in the value chain that occur at short notice. In Industry 4.0, companies' individual production lines organise themselves independently according to demand. If one machine on this line breaks down, production is reorganised autonomously via an alternative channel.
- Reduce lead times: Seamless data collection enables the rapid use of production-relevant data for near-term decision-making regardless of the location. This means Industry 4.0 users can reduce market lead times for innovations. Industry 4.0 in particular presents start-up firms with especially attractive options.
- Adapting to customer requirements with small batch sizes: Industry 4.0 allows the incorporation of individual customer-specific criteria concerning design, configuration, ordering, planning, production, and operation as well as enabling modifications to be made at short notice. Industry 4.0 is intended to ultimately even enable rapid and inexpensive low-volume production runs right down to one-offs (a batch size of 1); for example, in the automotive or furniture

manufacturing sectors (see Corò et al., 2017, p. 73). Smart production organisation and new technologies such as 3D printing processes are going to be used.
- New offerings of downstream services: In the business-to-business segment, Industry 4.0 opens up potential for high-performance services for the near-term evaluation of big data.
- Attractive work structuring: Modern organisational structures enable production that is more flexible. This chronological and spatial flexibility is likely to be attractive to many workers. This in turn is also a positive argument in a labour market that is changing due to demographic developments and given the shortage of skilled workers that is set to become increasingly acute. After all, a company with advanced, flexible work models should have the best prospects in the contest for increasingly scarce "top talents"; this applies even more if it also becomes involved in the design of the required further education and training measures (see Schlund et al., 2018, p. 278).

2.2.6 Challenges at technical, economic, legal, and organisational part

While there are highly promising benefits associated with Industry 4.0, it does also present several technical, legal, economic, and organisational challenges for companies in the value chain (see Chromjaková, 2017, p. 130). The most important of these challenges lie in using the collected data in real time, the utilisation of production capacities, the complexity of production organisation, integration in the value chain, as well as the issues of data protection and data security (see Prause and Günther, 2019, p. 42).

- The real-time evaluation and utilisation of the high volumes of data created in the value chain: The be-all and end-all of the Industry 4.0 concept is that data collected along the value chain is evaluated for its relevance in real time and utilised for organising production. This necessity represents one major specification for IT systems (see Soder, 2016, p. 21).
- Optimum capacity utilisation in autonomously organised production: Production in an Industry 4.0 company can deal flexibly, efficiently, and quickly with fluctuations in production and malfunctions in individual production areas using the available capacities. This optimisation procedure is extremely complex and will come into conflict with discretionary ad hoc interventions.
- Integration in the value chain: The Industry 4.0 idea is based on rapid and efficient data transmission within the value chain. This objective is predicated on infrastructures and processes being coordinated as

well as interfaces and protocols being clearly defined. As long as there is no general standard for these technical prerequisites, that is, each proprietary system is incompatible with the other, it is extremely costly for a company to switch into a different value chain; and in some cases it is actually economically impossible (see Baum, 2013, p. 41).

- Data protection and data security are crucial issues: Opinions are divided about comprehensive data exchange along the value chain. After all, the advantage of production flexibility is offset by the disadvantage of the possibility to close a link between supplier and customer and a disclosure of company processes (i.e. extremely sensitive internal strategic data). The risks pertaining to data protection and data security are serious: the opportunity to be able to adjust applications does make Industry 4.0 appealing to users. However, the extensive exchange of data that accompanies Industry 4.0 also makes users an attractive target for hackers. In such attacks, the intention may not only be the theft of relevant data, but also sabotaging the entire production process, so these may be risks with a macroeconomic dimension.
- A foretaste of the risk is the case of the Lukas hospital in Neuss/Germany. After an attack by a Trojan in 2016, the hospital was totally paralysed for several days – including the emergency units – and run up a bill of more than EUR 500,000. The research department of Deutscher Bundestag expects the collapse of civilian life in a worst-case scenario for massive hacker attacks (see Petermann et al., 2011, p. 233).
- Accordingly, when the contract is being drafted, the diverse facets of data protection and data security are extremely important to Industry 4.0 users. For example, it needs to be examined whether the exchange of critical data outside the EU (European Union) already contravenes the legal provisions that apply to the company. This applies in particular because other jurisdictions have completely different rules regarding data protection and data security, especially also regarding the scope for government access to the data.

2.2.7 *Expectations sometimes disappointed*

Some of the hopes generated by the initial euphoria for Industry 4.0 remain unfulfilled in practice. There is a great deal of uncertainty in this new field – and it is therefore imperative to query more closely the processes and steps required.

This means it is certainly possible that the total savings in a full cost calculation would actually be negligible for numerous firms. There is even the danger that the economic efficiency may decline overall, at least in the short term. Since those companies that choose Industry 4.0 initially have to make significant investments (e.g. for consulting, software, hardware, training, and fundamental reorganisation), it will be difficult to make a sound empirical estimate of the overall effect on costs and this will undoubtedly also take many years to be achieved.

2.3 Various business ideas

2.3.1 Market still in its infancy

The development of the Industry 4.0 market is still largely a vision of an evolutionary process that will not become truly tangible until the second half of the coming decade. This long-term process also intrinsically contains great uncertainties. These uncertainties are also expressed in the surveys and assessments presented by experts.

The current situation is typical of a new trend such as Industry 4.0. The idea is thus gaining ground much more slowly than had been widely hoped. This is also due to the opposition and structural obstacles that exist both on the part of users and on the part of providers of Industry 4.0 offerings (see Chromjaková, 2017, p. 131). This applies especially to the following four aspects:

- Opposition from the providers themselves: The comprehensive concept associated with Industry 4.0 forces providers in particular to reexamine their traditional business models. In numerous areas, the hitherto usual on-site provision of services, software, and hardware will be replaced. With these new offerings, the sales channels, value chains, and thus the margins will change markedly, as well. This is likely to encounter opposition within the organisations – at least in the early phase of the reorientation.
- Opposition from within the departments of the outsourcing companies: Before implementation of the restructuring measures, the user firms have to persuade the individual departments to cooperate in the execution of these changes. The objective of this persuasion is to ensure that unfounded fears do not prevent the introduction of new technologies and processes that are advisable for the company as a whole. These fears are likely to be wide-ranging – from losing one's job to having one's responsibilities curtailed (see Fergen, 2017, p. 225). Openly addressing all the concerns is thus likely to be a task that the management should not underestimate.
- The lack of generally applicable standards: A recognised technical standard could accelerate the universal interoperable use of fully automated offerings from the wide-ranging Industry 4.0 portfolio and overall market penetration; this is particularly true as numerous potential users are currently still holding back because of their uncertainty. These interested parties are particularly fearful of investing

considerable resources in an offering, which might not correspond with the standard that eventually becomes established. Reaching agreement on a generally applicable standard is, however, likely to prove difficult for a variety of reasons. On the one hand, there are certainly also discernible cultural differences between the companies in the mechanical engineering, electrical engineering, electronics, and IT sectors with regard to the multidisciplinary topic, especially also with regard to the path to standardisation. Standardisation in the fields of electrical engineering, electronics, and mechanical engineering, for example, has traditionally been more reliant on repeated time-consuming consultations and committee work than on IT. On the other hand, many providers could also make the strategic decision to rely on proprietary offerings. Adopting such an approach could be allied to the hope of using the restrictions on interoperability to make it much more difficult for customers to switch to a different provider.

- Enhancing network availability and speed: Communication networks must be able to handle the constantly rising volume of data and quality requirements. If the communications network has insufficient capacity (especially with regard to system availability and speed), then fully automated processes will be affected by interruptions to operating processes and high downtime costs will thus be incurred, as well. The prospects of success for the new automation offerings therefore depend directly on the reach and performance of the communication network. The issue of the commercial foundations for Industry 4.0 therefore is linked to the central macroeconomic issue of the expansion of the communications network.

2.3.2 *A wide range of entities …*

Experience clearly shows that Industry 4.0 offerings need to be tailored specifically to the company and cannot be supplied "off the shelf". So, the idea of Industry 4.0 can basically be conceived in very diverse contexts (see Ittermann and Niehaus, 2015, p. 41). The varied benefits associated with this development are shown by projects and prototypes developed by research establishments such as Deutsches Forschungszentrum für Künstliche Intelligenz (research institute for artificial intelligence, DFKI), or Fraunhofer ISS/EAS, Fraunhofer IOSB, and companies such as Agco (agricultural technology), BorgWarner (drive engineering), Brütsch-Rüegger (tool technology), Bosch Rexroth (drive engineering), Bruker (measurement engineering), Daimler (automotive), John Deere (agricultural technology), Festo (process technology), Harting (network technology), Homag (mechanical engineering), HP (IT), Introbest (systems engineering), Kaba (safety engineering), SAP (IT), Seca (measurement engineering), Sick (sensors), Siemens (electrical engineering), Stahlwille (tool technology), Trebing + Himstedt (process technology), Trumpf (manufacturing engineering), Volkswagen (automotive) or Wittenstein (propulsion technology) in the western hemisphere; and also Hisense (electrical engineering), Huawei (telecommunications), Lenovo (IT), Mindray (medical engineering), Shinva Medical (medical engineering), Spreadtrum (semiconductors), Xiaomi (smartphones), or ZTE (IT) in China (Figure 2.3).

2.3.3 *... and a wide range for applications*

The corresponding breadth of applications is indicated by the following examples:

- Remote maintenance reinvented: Currently the remote maintenance of highly specialised production machinery requires individual communications solutions. This is how the mechanical engineering specialist links up to the company network of the production facility. Configuring and managing this link-up is very costly. In the smart factory of the future, the production system links up with the appropriate experts via a telepresence platform depending on the situation. The machines autonomously supply the relevant data for the diagnosis.

Figure 2.3 Wide range of players in Industry 4.0 (selection). (Source: Author's illustration.)

- Changing supplier during the production process is made possible: At present, the manufacturer has a very big problem if a supplier is suddenly unable to deliver. It has to be quickly clarified how long the inventory will last, which products are affected and which firm could be the replacement supplier. Resolving such issues quickly enough is very complex and requires high-quality infrastructure. Industry 4.0 will directly enable the overall impact on value added to be estimated, including inventory level and logistical processing. Costs, the action required, and risks can thus be derived sufficiently quickly. The comprehensive linking-up of production systems enables alternative suppliers and their capacities to be analysed in real time. Enquiries can then be made to the alternative suppliers in a supplier cloud, for example.
- Plant engineering resources are to be used efficiently via practical shutdown capabilities: Many production facilities currently continue to operate even during extended manufacturing downtimes, such as non-working shifts and weekends, and thereby consume large amounts of energy. The plant remains in operation around the clock so that there is no need for a long start-up phase. During production-free periods energy is thus wasted senselessly, especially by robots, extraction devices, laser sources, and cooling equipment. Industry 4.0 enables this efficiency-enhancement potential to be tapped more effectively – assuming that the concept is already incorporated into the planning phase. With energy efficiency in mind, production-free periods for the machines are quickly recognised in order to switch them into an energy-saving standby mode. Demand-managed motors are used especially for this on extractor systems. Siemens has calculated that for a typical facility where laser welding technology is used for body-making, this instrument could cut energy consumption during rest periods by 90%, which would result in an overall energy saving of 12% (i.e. consumption would drop from 45,000 kWh to 40,000 kWh).
- End-to-end system engineering enables customised products to be made: Usually the value chain structure is highly path-dependent and also static in nature. The special requirements of individual customers have hitherto not been incorporated into the production planning process – even though this flexibility in the manufacturing process would in fact be possible in principle.

Industry 4.0 enables end-to-end engineering to be implemented. Production is monolithically developed from start to finish and thus coordinated with product development. Consequently, customers will be able to combine individual functions and components in the production process inexpensively according to their own requirements. Automaking is currently based on a predetermined production line. The functional scope of the individual stations is statically defined. The workers' tasks are geared directly to the requirements of the respective stations and are usually quite monotonous. If retooling is necessary in order to introduce a new product version, this is time consuming and thus also expensive. This means that individual customer preferences, such as the fitting of a component from a different product group from the same firm, cannot be considered.

The factory of the future consists of a flexible production line grid. The vehicle to be manufactured then manoeuvres autonomously through the production stations as a smart product. The flexibility of the production process allows a wide range of variation that is not limited by any central guideline.

- Agriculture is always also an engineering matter: Agriculture operates with biological systems, such as soil, plants, animals, and the weather, which at the abstract level are linked with similar risks to those we know from the industrial environment. In a similar way to the industrial sector, the agricultural sector also faces global competition. It is thus to be expected that the innovations from the fourth industrial revolution will also transform the agricultural sector. Going forward, making production more flexible and customising products will certainly become relevant in agriculture, too. Integrating customers into value chains (as entities demanding defined product features, such as food retailers who use their own quality standards to try to distinguish themselves from their competitors) is equally conceivable as a cyber-physical system as is the inexpensive creation of standard products. In pilot projects associated with precision farming, mechanical engineers and IT experts have joined forces with agricultural experts and are already testing sensor-based fertilising modelled on a cyber-physical system. This also still requires well-qualified personnel, that is, especially farmers with agro-technical expertise.

2.4 *China's way promising*

2.4.1 *Economical structures promote leapfrogging*

China's business environment is one of the most dynamic in the world. The labour market, in particular, has undergone dramatic changes and will continue to do so over the country's course of economic development. On the one hand, rapid economic growth has increased the demand for labour (Figure 2.4). On the other hand, China's seemingly infinite supply of labour is set out to shrink at the end of the decade, with an ageing population increasing the country's dependency ratio (age-population ratio of those typically not in the labour force to those in the labour force) (Figure 2.5).

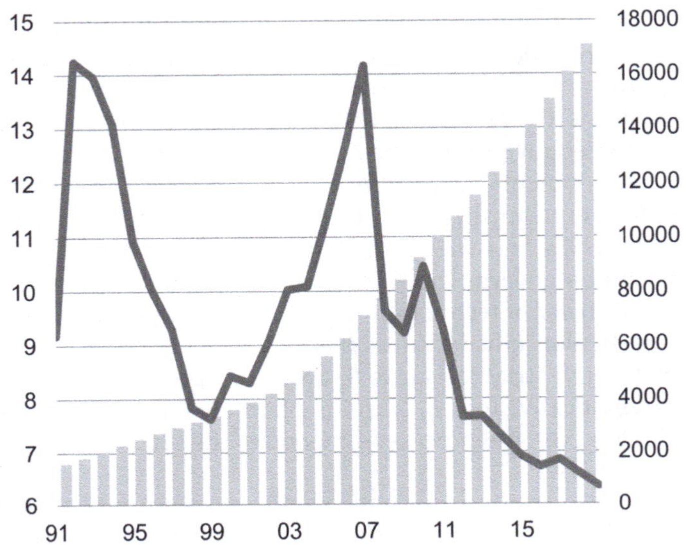

Figure 2.4 Rapid economic growth in China: gross domestic product (GDP) per capita, USD in purchasing power parities (right); percentage year over year (left), China. (Source: National Bureau of Statistics of China, 2019.)

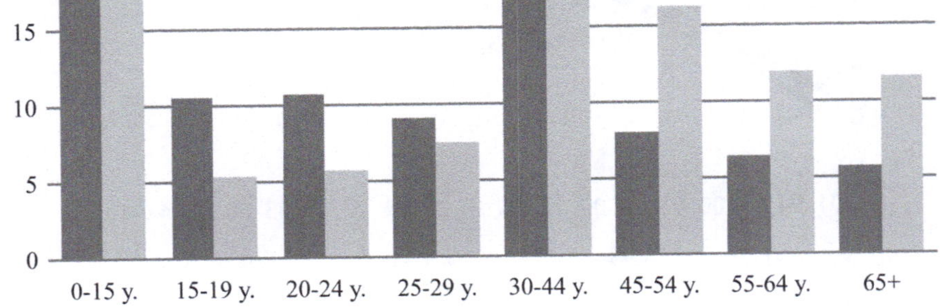

Figure 2.5 China's aging population: percentage by cohort in China. (Source: World Bank, 2019.)

Wages respond to the change in demand and supply of labour, giving rise to a new upper middle income consumer class. In consequence, the country's comparative advantage as well as its domestic demand structure shifts towards more capital and technology-rich end-user commodities.

The Chinese government has responded early to the challenge. Economic development shifted from a labour-intensive approach in agriculture and manufacturing towards capital-deepening industrialisation from the mid-1990s onwards. Besides this, not least due to the growth recession, China's government facilitated improvements in cost innovation and the transfer of technology through inward foreign direct investments, in particular joint ventures (Han et al., 2016, p. 139). As a result, the economy's industrial structure as well as export basket composition changed dramatically. Today, China is often labelled "the workshop of the world". However, the process of economic transformation is unlikely to end there.

The next step on the transformation ladder does not only involve a shift into higher-value added manufacturing, but also moving from "Made in China" to "Designed in China". This will be key in ensuring international competitiveness in an ever-changing global market. More specifically, it requires embracing the huge benefits associated with Industry 4.0 by actively promoting its comprehensive implementation within the business environment (Figure 2.6).

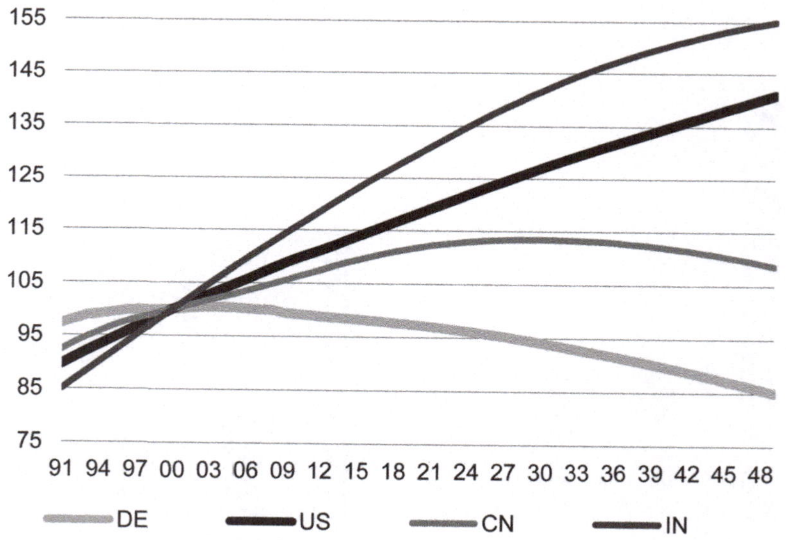

Figure 2.6 Declining population in China and Germany ahead: population by number, Index 2000 = 100. (Source: OECD, Statistisches Bundesamt, own calculations.)

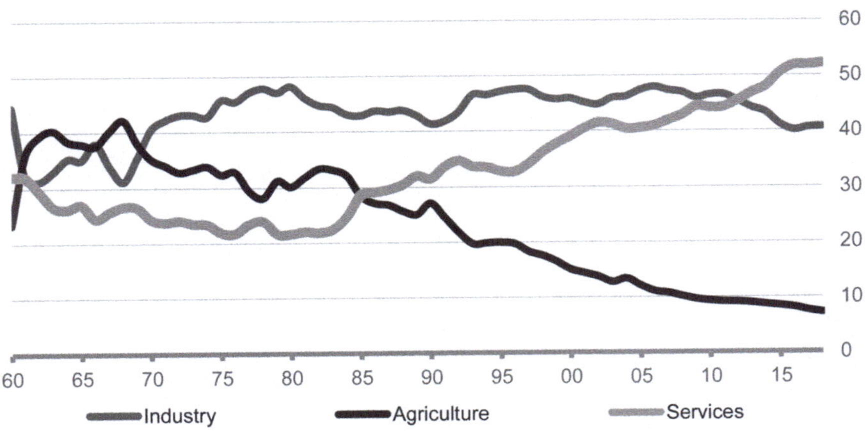

Figure 2.7 Rapid structural transformation: value added as percentage of total gross domestic product (GDP), by sector, China. (Source: World Bank, 2019.)

2.4.2 *Huge potential waiting to be harvested*

The potential gains for China are immense, as they scale with the size of the industry. The country is the world's largest manufacturer (Figure 2.7). The output value of China's equipment manufacturing industry alone surpassed USD 3.2 trillion in 2013, which accounts for around one third of the global output. China's Ministry of Industry and Information Technology estimates that the economy's automation industry is worth around USD 100 billion. According to the Industrial Federation of Robotics, the Asia-Pacific region is already the main market for robots. Because of the geographical position alone, for ground-breaking Chinese firms this is a major opportunity to gain market shares. The boost in labour productivity becomes more and more a necessity to compensate for the loss in labour cost advantages and to ensure international competitiveness (see Das and N'Diaye, 2013, p. 16) (Figure 2.8).

What is needed to successfully implement Industry 4.0? The good news is that, in principle, the technology to realise this vision is available to us today. It comprises production and performance reporting, condition-based monitoring, intelligent alerts and event management, calculation of key performance indicators, and, ultimately, technology that promotes collaboration for improved decision-making. The necessary precondition for utilising this technology is a modern business environment defined by a variety of capabilities, such as high technological capacity of the workforce and a well-established ICT infrastructure (see Mertens, 2014, p. 29).

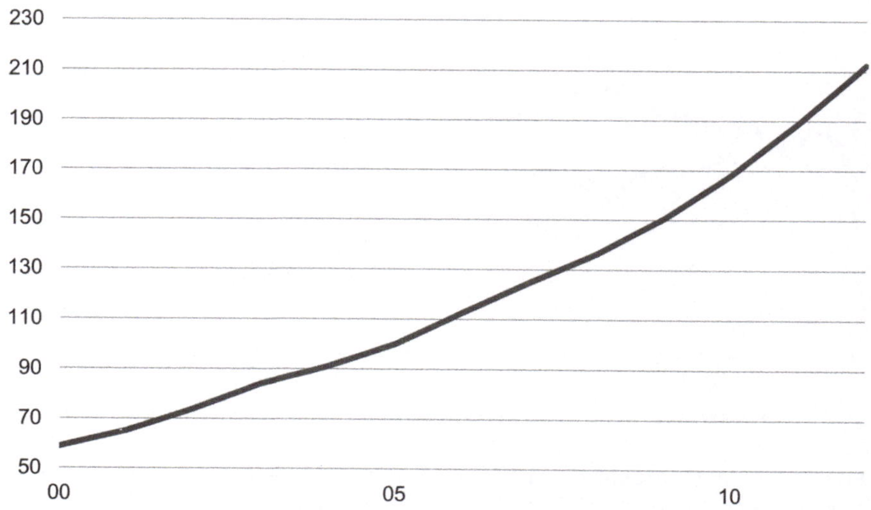

Figure 2.8 Wages in manufacturing rise steadily: real wages in manufacturing, China, 2005 = 100. (Source: World Bank, 2019.)

For them to be in place, awareness is the starting point. The Chinese government has explicitly emphasised the importance of automation in the context of the Internet of Things and services for China's transition. Moreover, politics started to lay the foundation for the successful adoption and incorporation of Industry 4.0 into the existing industrial framework by means of various policies.

2.4.3 *Impulses laid by politics*

China's policy approach to Industry 4.0 can be divided into three pillars. The first concerns the state's role in the creation of an appropriate environment in which industrialisation can take place. The second concerns the state's direct intervention in the process of industrialisation. Finally, the third pillar aims to gradually relocate the source of innovation from public institutes to private companies.

Several policy examples can be stated. For instance, 15 years ago, China's State Council decided to accelerate the economy's technological progress via innovation, high-technology adaption, and industrialisation by issuing the "Circular on policies for the development of software and IC industries". The program consists of policies spanning across all fields, including investment, financing, taxation, industrial technology, exports, income distribution, training, government procurement, and intellectual property. China's industrial transformation was further facilitated in 2011

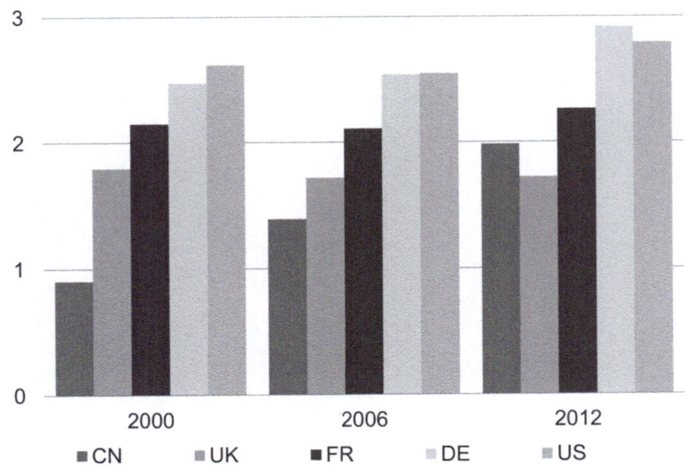

Figure 2.9 China's increased spending on research and development (R&D) expenditure as percentage of gross domestic product (GDP). (Source: World Bank, 2019.)

with the "Circular on the issuance of further encouraging the development of software and IC industry" (Figure 2.9).

In 2012, the 18th Central Committee designated ICT-oriented development as one of China's key societal and economic goals until 2020. With the 12th Five-Year Plan, high-tech companies are supposed to become main drivers of China's development. Specifically, the government highlights seven priority industrial segments. These sectors are new IT, high-end equipment manufacturing, biotechnology, energy conservation, new energy, new materials, and clean-energy vehicles (see Nuernberg and Wang, 2012, p. 109). To put this into perspective, the budget of the EU "Horizon 2020" program amounts to around EUR 80 billion for 2014 to 2020 (Figure 2.10).

In 2015, China's State Council issued the "Made in China 2025" plan. The 10-year plan, which is inspired by Germany's "Industry 4.0" plan, has been compiled by more than 20 departments, 50 members of the Chinese Academy of Engineering, and 100 other experts. It sets out to make innovation the core competitiveness of an intelligent, digital, and networked Chinese manufacturing industry and comprises the construction of a national manufacturing industry innovation centre. The plan reinforces the importance of strategic emerging industries, highlighting a slightly modified and extended list of ten priority sectors, which are as follows: new advanced information technology, automated machine tools and robotics, aerospace and aeronautical equipment, maritime equipment and high-tech shipping, modern rail transport equipment, new-energy

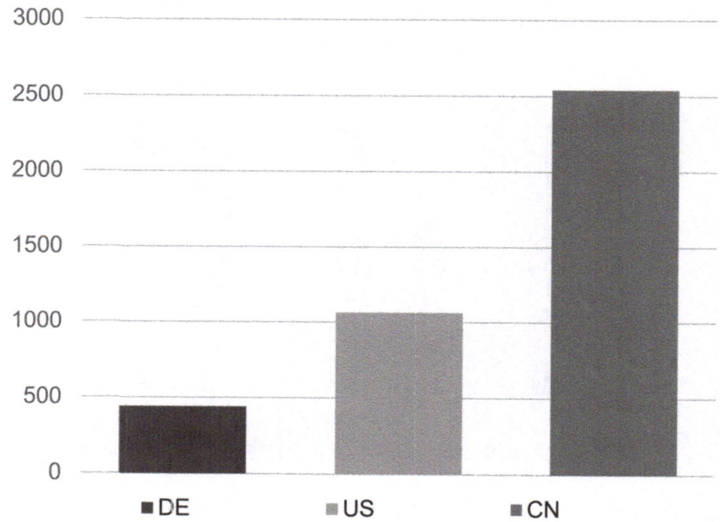

Figure 2.10 China leads in patent applications: new priority patent application in the field of Industry 4.0. (Source: Anon. 2015, p. 8.)

vehicles and equipment, power equipment, agricultural equipment, new materials as well as biopharma, and advanced medical products. However, opposed to industrial five-year plans, the strategy "Made in China 2025" puts emphasis on a comprehensive upgrading of the economy's industrial structure.

2.4.4 Policies leads to first outcomes

Today, the outcomes of China's past policy actions are sizeable. According to McKinsey, the management consultancy, hundreds of mid-sized companies are creating a Chinese version of the German Mittelstand, with a focus on biotech, pharmaceuticals, consumer electronics, medical technology, drones, and telecommunications equipment (see Orr, 2014, p. 11). And they are getting more and more successful to become global market leader in their niches (Anon., 2018, p. 5). The relative size of China's internet economy has not only caught up by international standards, but its share of 4.4% of gross domestic product (GDP) in 2013 also ranks the country among the global leaders, surpassing advanced economies such as the United States, France, and Germany (see Woetzel et al., 2014, p. 25).

Improvements have been made on various levels, ranging from the formation of worker's technical skill capacity to patent applications and infrastructure development. More specific, the country surpassed the

United States and Germany in terms of new priority applications for patents in the field of Industry 4.0. China is now also the economy with the largest number of machine-to-machine connections (50 m), followed by the United States (32 m) and Japan (9.3 m) (see GSMA, 2014, p. 7). In terms of expenditures in research and development (R&D), the country is on its way to reach its target to spend annually 2.5% of its GDP on R&D by 2020. The steady increase in the R&D expenditure share is particularly impressive given the large GDP growth rates in the last decade. In light of these numbers, it is not completely surprising that 57% of the executives interviewed in a survey published by Infosys (2015), a business and IT consultancy, stated China to be the most mature adopter of Industry 4.0, ahead of the United States (32%), United Kingdom (26%), Germany (21%), and France (14%).

2.4.5 Highly heterogeneous regions

However, these numbers do not describe China of today completely. Clearly, China's manufacturing industry has made great achievements in fields like high-speed rail transportation equipment and telecommunication facilities. Companies such as Hisense (electronic hardware), Huawei (telecoms equipment), Lenovo (IT hardware), Mindray (medical equipment), Shinva Medical (medical equipment), Spreadtrum (semiconductor equipment), Xiaomi (smartphones), and ZTE (IT hardware) are among the world's leading innovators in their field, employing myriads of white-collar workers in their R&D departments.

At the same time, however, the majority of Chinese firms have not yet fully embraced the innovations related to past industrial stages. Many local firms are either reluctant or lack the financial capacity to invest in automation (see Terminio, 2015). McKinsey estimates that the internet adoption ratio of China's SMEs reaches only 25% (see Woetzel et al., 2014). This means that, overall, digital unawareness is still particularly pronounced in this large group of companies, which together generate roughly 70% of China's GDP (Figure 2.11).

In addition, the reality is that most of the market leaders in the field of robotics and automation that produce in China are foreign companies, as for instance ABB, Comau, Fanuc, and Yaskawa (see Terminio, 2015). These firms generally form clusters in tech areas. These clusters are characterised by a well-established transport and ICT infrastructure. According to the Hong Kong Trade Development Council, there are more than 260 tech cluster in China today. Such cluster we find, for example, near Beijing, Chongqing, Shanghai, Tianjin, or in particular in Shenzhen in the Hong Kong area. These agglomerations are extremely dynamic and innovative. Thus, in Shenzhen the average age of the 15 million inhabitants (for comparison, Hong Kong: 7 m) is 27 years (for comparison, Berlin: 43 years).

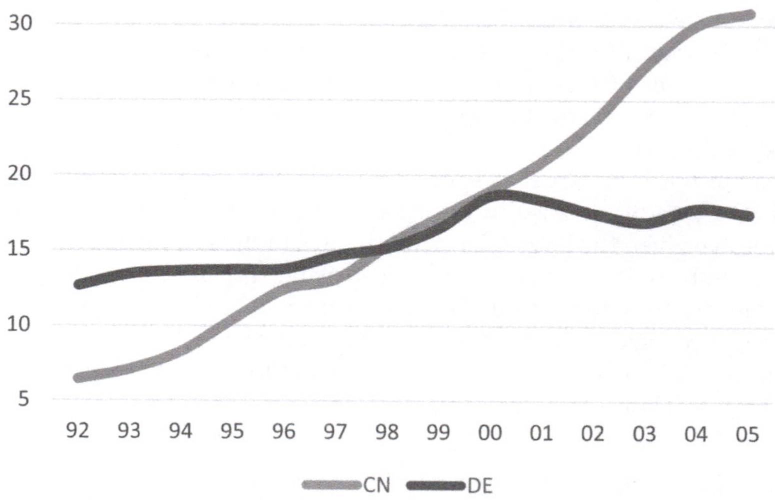

Figure 2.11 China's high-tech exports increase in relative importance: high-technology exports,[3] percentage of total manufactured exports. (Source: World Bank, 2019.)

Besides, 15% of all unicorns[2] worldwide, or world market leader like WeChat (internet communication and payments), Tencent (gaming), DJI (drones), or BYD (e-cars), are located here – not in the directly nearby city Hong Kong. Finally, all of 21,000 taxis and all of 16,000 urban buses are electrically driven.

Nevertheless, there is considerable heterogeneity between China's provinces, industries, companies, and the workforce in terms of readiness and use of technology. Therefore, Industry 4.0 realises locally and in an evolutionary fashion: factory by factory and company by company. Clearly, this makes it rather difficult to make a clear statement on the state of Industry 4.0 in China, as the answer needs to be: "it depends". It depends on whether one looks at a single factory, company, value chain, industry cluster, or the economy at large.

However, it is also evident that in order to repeat fully the benefits of Industry 4.0, its implementation cannot be limited to a few companies. The gains to each company increase with the number of integrated industries and actors. To put it differently, in the context of Industry 4.0, the total is larger than the sum of its parts – exponentially larger. From this

[2] Start-ups with a worth of more than USD 1 billion.

[3] High-technology exports are products with high R&D intensity, such as aerospace, computers, pharmaceuticals, scientific instruments, and electrical machinery.

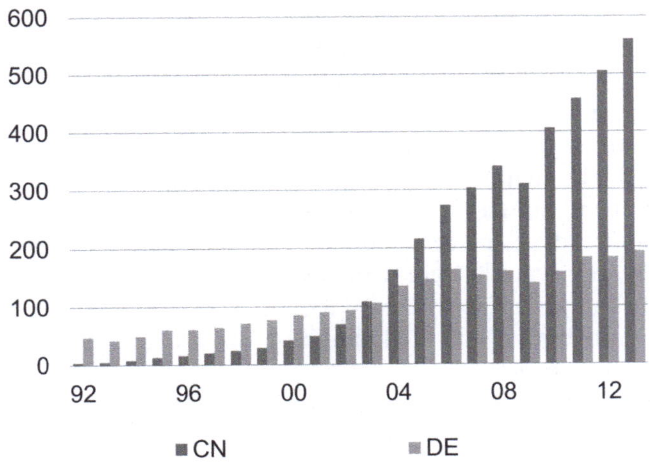

Figure 2.12 China's high-tech exports increase in large absolute values high-technology exports, USD billion. (Source: World Bank, 2019.)

standpoint, China's implementation of Industry 4.0 is still in its infancy. Technological innovation and industrial upgrading have yet to trickle down to realise the vision of an innovation- and knowledge-driven economy. A long road lies ahead (Figure 2.12).

2.4.6 *Enable and facilitate company-level action*

A vital component of the digital evolution is a rapid and efficient data transmission within the value chain. This requires coordinated actions among various companies. However, the implementation and innovation of Industry 4.0 at the company level are heavily influenced by the availability and accessibility of complementary (public) inputs. Clearly, a company can be constrained on its path to industrial upgrading by a lack of available advanced power grids and broadband connections. Unfortunately, complementarity creates problems of attribution. The question of who has to bear the costs for the provision of needed inputs will be closely linked to the transformation process.

A market revolution is prone to be accompanied by market failures and often calls for state action. For instance, a key challenge of Industry 4.0 lies in the productive use of collected data in real time. However, no company will bear the costs of countrywide high-speed broadband coverage. Companies are also reluctant to invest heavily in the quality of the workforce given the possibility of inter-company migration. At the same time, companies will lack incentives to build up other company capabilities needed to incorporate Industry 4.0 into their production process, if either the ICT infrastructure or the skill-biased demand for labour is not met.

Unfortunately, there is no standard blueprint for the rollout of Industry 4.0. The shape of concrete actions needed in terms of infrastructure investment, source of funding, and time horizon depends on the conditions already in place. In an environment where the above-stated constraints are severe, a hybrid policy model is suited. Innovation will come from bottom-up approaches, while top-down policies address the problem of attribution and facilitate the process of industrial upgrading on the company level.

That said, the strategy "Made in China 2025" is based on four principles, which together indicate a reorientation in China's policy approach. These principles are market-oriented and government-guided, based on the present and having a long-term perspective, comprehensively pressing forward, and making breakthroughs in key areas, independent development, and win-win cooperation, according to Miao Wei, former Minister of the Ministry of Industry and Information Technology.

Hence, compared to the current industrial Five-Year Plan, "Made in China 2025" means a shift away from a rather narrow focus on selected strategic emerging industries. The importance of the overall development of China's economic society is recognised and highlighted. Beside this, the role of market institutions is stressed. Following the statements of highly ranked Chinese officials, this includes the dedication to create a fair environment that stimulates innovation and opening up much wider in terms of scientific and technological research from foreign initiatives. Explicitly, it highlights the dominant role of enterprises in technology innovation.

Next, despite the top-down policy approach, China seems to understand that cooperation is the key to complete the steps to an innovation-driven, high-technology growth model that implements Industry 4.0. Domestically, a council was established in 2013 to coordinate the government's policy and action on the Internet of Things. Importantly, the council's members include various public institutions, including the National Development and Reform Commission, the Ministry of Industry and Information Technology, the Ministry of Science and Technology, the Ministry of Education, and the National Standardization Administration. However, also Chinese executives consider their voice heard by politics and regard a range of national public policies to contribute to their competitive advantage, according to a global CEO survey by Deloitten, an advisory firm, and the U.S. Council on Competitiveness (2013). This seems to be less the case in Europe and the United States.

2.4.7 Improving infrastructure

The ability to set up and use the latest available technology is a key factor for a comprehensive upgrading of an economy's industry, and ultimately for the implementation of Industry 4.0. According to the Global Competitiveness Report (GCR) of the World Economic Forum (WEF), an international institution for public-private cooperation, China has still much room for improvements in this regard. While the number of people connected to the internet has risen considerably in the past, internet adoption has been mostly consumer-driven. Moreover, regarding internet bandwidth on average, China is still lagging behind (Figure 2.13).

Nonetheless, based on the policy targets, the government is determined to change the status quo (see Jia et al., 2019, p. 14). The prerequisite of a modern ICT infrastructure is addressed by the National Broadband Plan. The plan aims to provide internet coverage at speeds of 20 megabits per second (Mbit/s) in urban areas and 4 Mbit/s in rural areas. China's government plans to spend USD 603 billion in the machine-to-machine ecosystem alone until the year 2020 (see Anon., 2014, p. 7). To put this into perspective, the German government targets to connect every German household with broadband internet (50 Mbit/s) based on fibre-optic technology until 2018. The TÜV Rheinland (2013) calculates that the costs will amount to around USD 100 billion.

It remains to be seen how ICT investments are regionally prioritised. Since the 12th Five-Year Plan, investments in strategic emerging industries are linked directly to a focus on single provinces. For example, the

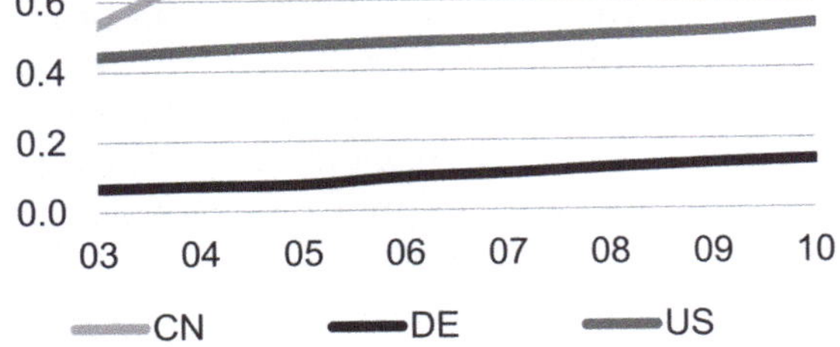

Figure 2.13 Rapid academic increase: number first university degrees in science and engineering. (Source: National Science Foundation, 2014.)

development of an industry for "clean energy vehicles" is concentrated in Beijing, Shanghai, and Anhui, while the "new material" industry is focused on Shaanxi, Hunan, and Liaoning (see Nuernberg and Wang, 2012, p. 21). Although the "Made in China 2025" plan takes a more comprehensive approach, a sector focus remains. Hence, it is likely that public infrastructure investments will continue to foremost further upgrade regional industrial cluster.

2.4.8 *Challenging the quality of the workforce*

Talent-driven innovation is considered the most important driver in the Global Manufacturing Competitiveness Ranking (see Giffi et al., 2016, p. 22). Gearing people up with the skills they need to conduct the tasks demanded, involves improvements on various levels. It means boosting the ability to adopt an interdisciplinary mindset, continuing to foster technical and analytical skill creation, as well as enabling and promoting lifelong learning. This can be achieved by improvements in the quantity and quality of formal education, as well as the availability of basic and advanced vocational training.

China presents a mixed picture in this regard. Undoubtedly, progress has been made in terms of school enrolment rates. Today, the enrolment rate in secondary education is about 90%, coming up from 58% in 2000.[4] Regarding tertiary education, the enrolment rate was only 8% in 2000 and increased to about one third today. More specific, the number of graduates achieving a first university degree in the fields of science and engineering grew by one sixth p.a. since 2000. Correspondingly, China awarded about 320,000 bachelor degrees in natural sciences and about 820,000 in engineering.

However, progress comes from a low level and the large absolute numbers of bachelor degrees in areas such as natural sciences and engineering mask the low average skill capacity per worker considering the country's enormous total workforce. Despite the improvements, the country still lacks behind in international comparison. In the GCR of the WEF, China is only ranked 72nd and 85th in terms of secondary and tertiary education enrolment rates. In addition, no progress has been made between 2000 and 2010 regarding the share of people in the age of 15 and above that completed tertiary education, which remains on a low level of 2.7%. Hence, the level of education of the workforce currently employed is still rather low. Further improvements in society-wide education are needed to avoid a severe shortage of skilled labour and to

[4] The gross enrolment rates of schooling is the ratio of the total number of enrolled student in a certain level (primary, secondary, or tertiary) divided by the population in the respective age group (e.g. school age or college age).

increase China's capacity to trickle down, adopt, as well as invent new technologies on a wide scale.

2.4.9 Constructively addressing existing opposition

Chinese assembly lines today differ greatly from those of a decade ago. Over the years, the cost of technology that replaces labour in factories has plummeted, displacing more and more workers. As it is often the case, it is the most vulnerable sections of society that bear most of the costs. Automated processes will replace routine-based jobs while augmenting analytical and technical skilled labour. However, it will also generate economic growth in aggregate terms.

Numbers might help to address general opposition: McKinsey estimates that companies which adopted internet technologies created 2.6 jobs for each job lost (see Woetzel et al., 2014, p. 24). In China, almost 1.5 million people have taken employment in express delivery and shipping businesses with the help of the rapidly growing e-commerce. Hence, Industry 4.0 will enlarge the overall pie while, however, also affecting its distribution (Figure 2.14).

Addressing a potential shortfall in workers' capabilities via specific policy action has therefore two key benefits. On the one hand, it raises the quality and mobility of the workforce, and hence the awareness and

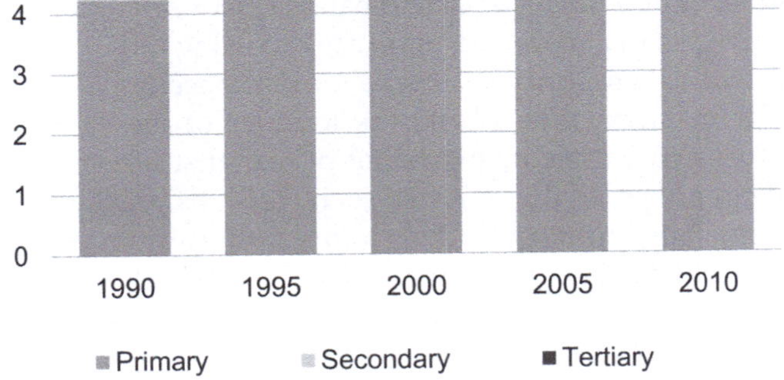

Figure 2.14 Almost no progress in tertiary education: average years of schooling, by education level, China. (Source: Barro and Lee, 2013, p. 189.)

support for Industry 4.0 on an individual level. On the other hand, it motivates companies to invest in capabilities that promote the implementation of Industry 4.0, which boosts efficiency and aggregate economic prosperity.

2.4.10 *Ensure legal framework of data security and data protection*

To facilitate innovations and information flows, currently unresolved issues concerning data security, confidentiality, and standardisation need to be addressed. Interoperability and compatibility are essential for automation.[5] Efficient data transmission requires that interfaces and protocols are clearly defined. As long as different proprietary systems are incompatible with each other, it is extremely costly for a company to switch into a different value chain; and in some cases, it is actually economically impossible. Standardisation is therefore a technical prerequisite.

The risks pertaining to data protection and data security are serious and constitute a concrete obstacle to the implementation of Industry 4.0. Companies might abstain from a comprehensive integration into a value chain for fear of disclosing sensitive internal company data, if the links between supplier and customer become too close. More generally, the vast quantities of data uploaded and exchanged make users attractive targets for external attacks by hackers. Due to the increasing connectedness, such hacker attacks might then sabotage the entire production process, extending the level of risk from a company level to a macroeconomic dimension.

A potential conflict arises between the national or regional character of the prevailing jurisdiction and the internationality of value chains. The concern that intellectual property becomes disclosed is fuelled by the fact that there are currently major geographical differences between legal provisions in terms of data protection and data security, in particular in the scope of data access granted to the government. The U.S. Patriot Act as well as the Great Firewall in China provide two practical and explicitly formulated examples of challenges to balance between national interests and privacy rights in a more and more digitalised world.

The lack of data security can make the existence of patent protection obsolete. For instance, Elon Musk, the founder of the rocket company SpaceX, refuses patent technologies out of fear that foreign space agencies would simply pinch them. While such concerns prevail especially in some economically, military, and therefore politically important industries, they clearly display the dilemma of an increasingly integrated, digitalised, flat world.

[5] Interoperability is the ability of differing systems, technologies, or organisations to work together. This usually requires compliance with common standards. If two systems can join, they can be described as being compatible.

The "Made in China 2025" strategy calls for strengthening intellectual property right protection for SMEs and stresses a more effective use of intellectual property in business strategy. Moreover, the plan discharges a top-down approach in terms of setting unique domestic standards and instead focuses on self-declared standards and the international standards system. This is a step in the right direction. It remains to be seen how the rhetoric is put into action.

Inconsistent regulatory interpretation and unclear laws are ranked the second biggest business challenge in China according to the "China Business Climate Survey Report" published by the American Chamber of Commerce in the People's Republic of China (AmCham China). Although 86% of the surveyed AmCham China member companies state that intellectual property law enforcement has improved, 78% also consider it to still be ineffective.

Further efforts are needed, so that a high degree of trustworthiness in the legal framework prevails and company property, tangible or intangible, is considered to be secure. Discrepancies between respective legal frameworks might otherwise constrain international trade and investment flows. Clearly, this cannot be achieved without close international cooperation between governments and institutions.

2.4.11 *International cooperation on a level playing field*

Aside from domestic policies, international cooperation plays a crucial part in the digital transformation of industries. It is encouraging to see that China has made a range of efforts to foster international agreements, despite the intense competition on the firm-level. For instance, the China-Germany Standardization Cooperation Commission deserves positive recognition. The commission is staffed with 40 representatives from the Chinese and German governments, DIN (German Institute for Standardization), Standardization Administration of the People's Republic (SAC), and further associations including acatech (Germany's National Academy of Science and Engineering) and Plattform Industrie 4.0 (a consortium promoted by the associations for electrical engineering (ZVEI), mechanical engineering (VDMA), and the ICT industry (Bitkom)).

A different example is the establishment of the first general standard in the context of the Internet of Things, initiated by China and passed by the International Telecommunication Union (ITU) in 2012. Moreover, the new standard was followed up appropriately by a joint White Paper issued by the EU and China last year in which the challenges for future developments are presented and various directions and development guidelines are discussed.

A crucial ingredient for Industry 4.0 is to promote a bidirectional technology and knowledge transfer, which is mutually beneficial for all partners. China and Germany have laid a solid foundation in this regard. In 2014, China and Germany released their joint framework for action called "Innovation gemeinsam gestalten". This framework stresses Industry 4.0 as a central point for future bilateral economic relations.

The China-Germany "Industry 4.0 dialogue" and the "Deutsch-Chinesische Allianz für Berufsbildung" are great examples of how international cooperation can promote Industry 4.0 in both countries even in areas such as workforce quality enhancement. The initiatives aim to promote the joint education of skilled engineering workers and therefore directly address the implications associated with Industry 4.0.

2.5 *Conclusion: Innovation creating a buzz*

The concept of Industry 4.0. inhabits the wide-ranging terrain occupied by augmented reality, Big Data, cloud computing, cyber-physical systems, RFID chips, Internet of Things and services, machine-to-machine communication, precision farming, and Smart X. At present, the offerings associated with the Industry 4.0 buzzword are definitely suffering from the excessive marketing-driven expectations and the lack of a clear definition. It is therefore certainly possible that following the hype which typically surrounds such new ideas and the subsequent disillusionment, there will be no one talking about Industry 4.0 in a few years' time. All the same, the basic idea behind the buzzword has a good chance of being implemented – regardless of the term used to describe it.

The application of Industry 4.0 has largely been a matter for big companies. However, small- and medium-sized companies are also finding that vertical and horizontal integration are becoming increasingly important factors in the competitive international arena. The wide range of benefits associated with Industry 4.0 are being demonstrated by highly promising projects and prototypes from research institutes, such as DFKI, or Fraunhofer ISS/EAS, Fraunhofer IOSB, and companies such as Agco and BorgWarner via Homag, HP, and Introbest throughout Trumpf, Volkswagen, and Wittenstein in the western hemisphere; and also from Hisense and throughout Spreadtrum and ZTE in China.

The idea behind the buzzword Industry 4.0 has good prospects that extend beyond the coming decade. With the international web of trade flows continuing to intensify, automation, more flexible processes, as well as horizontal and vertical integration will become increasingly important for a competitive, modern manufacturing structure. This will apply even more if the currently still unresolved issues concerning control responsibility, security, confidentiality, standardisation, legal framework, and infrastructure configuration (e.g. expansion of advanced power grids and communications networks) are addressed constructively.

Here, China is determined to seize the outstanding opportunity at hand. So far, not least due to the growth recession, China's government has laid a foundation by means of various policies. On first sight, the country appears to be in a good position in the race of the digital evolution.

Hence, the country has set up world-leading high-technology manufacturers in the recent past and an emerging, highly innovative group of SMEs might actively participate in the digitalisation of value chains. However, the great majority of companies still has to implement features related to past industrial stages. Many companies and workers are either unprepared or even unaware concerning the digital evolution. For a comprehensive upgrade of China's economic structure and a thorough implementation of Industry 4.0 promoting and integrating SMEs is a vital component. Only then will the country be able to sustain its transformation process and truly grasp the enormous benefits of a digitalised economy.

For a comprehensive upgrading and digital transformation of the Chinese industry, further improvements in workforce quality and ICT infrastructure are urgently needed. With its "Made in China 2025" plan, China maintains a top-down oriented hybrid (or dual-track) industrial policy approach to transform its economy through market forces as well as strong government interventions. However, overall, the new strategy leaves more room for company initiatives and further opens the door for foreign actors to participate on a level playing field.

As the benefits of the digital transformation of the industry are likely to be not fully tangible until the next decade, patience is required. This ensures that actions are not only future-oriented but also linked to a healthy portion of pragmatism. With the current plans, which combine a long-term vision with concrete actions in the present, the Chinese government is on the right track for upgrading the entire value chain.

References

Ahrens, D. and G. Spöttl. 2015. Industrie 4.0 und Herausforderungen für die Qualifizierung von Fachkräften. In: Digitalisierung industrieller Arbeit, eds. H. Hirsch-Kreinsen, P. Ittermann and J. Niehaus, 185–200. Baden-Baden: Nomos.

Albrecht, T. 2016. An der Schnittstelle. In: Werkheft 01. Digitalisierung der Arbeitswelt, eds. A. Ammermüller, P. Kalkhake and H. Zirden, 100–103. Berlin: BMAS Abteilung Grundsatzfragen des Sozialstaats.

Anon. 2013a. Umsetzungsempfehlungen für das Zukunftsprojekt Industrie 4.0, Abschlussbericht des Arbeitskreises Industrie 4.0. acatech. Berlin.

Anon. 2013b. Revolution in vier Stufen. In: Industrie 4.0. Wenn das Werkstück die Fabrik steuert. Ampere – Das Magazin der Elektroindustrie, ed. ZVEI, 28–29. Frankfurt.

Anon. 2014. Connected Living. How China is set for global M2M Leadership. GSMA Report. London.

Anon. 2015. Analyse der Entwicklung von Industrie 4.0 in China. White Paper 1: Analyse chinesischer Patentaktivitäten, ed. Fraunhofer IAO, IAT, University of Stuttgart. Stuttgart.

Anon. 2018. Hidden Champions Top in Digitalisierung. Simon Kucher & Partners. Bonn.

Anon. 2019. China Business Climate Survey. American Chamber of Commerce in the People's Republic of China. Beijing.

Barro, R.J. and J.W. Lee. 2013. A new data set of educational attainment in the world, 1950-2010. Journal of Development Economics, 104(C): 184–198.

Bauer, W., et al. 2014. Industrie 4.0 – Volkswirtschaftliches Potenzial für Deutschland. Stuttgart.

Bauernhansl, T. 2014. Die Vierte industrielle Revolution – Der Weg in ein wertschaffendes Produktionsparadigma. In: Industrie 4.0 in Produktion, Automatisierung und Logistik, eds. T. Bauernhansl, M. ten Hompel and B. Vogel-Heuser, 5–35. Wiesbaden: Springer.

Baum, G. 2013. Innovation als Basis der nächsten Industrierevolution. In: Industrie 4.0 – Beherrschung der industriellen Komplexität mit SysLM, ed. U. Sendler, 37–53. Berlin: Springer.

Brynjolfsson, E. and A. McAfee. 2014. The Second Machine Age – Work, Progress, and Prosperity in a Time of Brilliant Technologies. New York, NY: W. W. Norton & Company.

Chromjaková, F. 2017. Stability of Business Process – Core Assumption for Implementation of Industry 4.0 Concept in Industrial Enterprise. In: Proceedings of the 4th International Conference on Finance and Economics, eds. D. Pavelková and P. T. M. Ly, 128–139. Ho Chi Minh City, Vietnam.

Corò, G., et al. 2017. Enabling Factors in Firms Adoption of New Digital Technologies – An Empirical Inquiry on a Manufacturing Region. Venice: Ca' Foscari University.

Das, M. and P. N'Diaye. 2013. Chronicle of a Decline Foretold: Has China Reached the Lewis Turning Point? IMF Working Paper 13/26.

De Propris, L. and D. Bailey. 2018. Industry 4.0: New industrial spaces and implications for industrial strategy. In: What we really mean when we talk about industrial strategy, ed. C. Berry, S. 46–52. Manchester: Manchester Metropolitan University.

Ezell, S. J., R. D. Atkinson, I. Kim and J. Cho. 2018. Manufacturing Digitalization: Extent of Adoption and Recommendations for Increasing Penetration in Korea and the U.S. Washington. Available at https://ssrn.com/abstract=3264125 or http://dx.doi.org/10.2139/ssrn.3264125

Fergen, A. 2017. Arbeitsschutz 4.0 – Essentials einer digitalen Humanisierungs-Agenda aus Sicht der IG Metall. In: Prävention 4.0: Analysen und Handlungsempfehlungen für eine produktive und gesunde Arbeit 4.0, eds. O. Cernavin, W. Schröter and S. Strowasser, 221–234. Berlin.

Giffi, C.A., M.D. Rodriguez, B. Gangula, A.V. Roth and T. Hanley. 2016. Global Manufacturing Competitiveness Index. Deloitte. London.

Gornig, M. and A. Schiersch. 2015. Perspektiven der Industrie in Deutschland. In: Vierteljahreshefte zur Wirtschaftsforschung, 1/2015 ed., 37–54. Berlin: Deutsches Institut für Wirtschaftsforschung.

Gushima, K. and T. Nakajima. 2017. A Design Space for Virtuality-Introduced Internet of Things. Future Internet, 9(4). https://doi.org/10.3390/fi9040060

Han, J., et al. 2016. Artificial Intelligence and Industry 5.0. Artificial Intelligence and Robotics Research, 6(4): 135–140.

Heininger, R. 2018. Empfehlungen für den Umgang mit Heterogenität in IT-Servicewertschöpfungsnetzwerken. München.

Heng, S. 2015. Industry 4.0: Outstanding Opportunity for an Emerging Market – But Patience and Cooperation Will Be Required. China Policy Review, 6(15): 10–13.

Heng, S. 2019a. Industrie 4.0. WISU: 567-574. Düsseldorf.

Heng, S. 2019b. Virtuelle Realitäten: Eine enorme Chance für das Management. In: Die Zukunft der Managementbildung, ed. F. Hubert, 33–35. Berlin: Jubiläumsschrift, Deutsche Akademie für Management.

Infosys. 2015 Industry 4.0: The state of the nations. Bengaluru, India: Infosys.

Ittermann, P. and J. Niehaus. 2015. Industrie 4.0 und Wandel von Industriearbeit – Überblick über Forschungsstand und Trendbestimmungen. In: Digitalisierung industrieller Arbeit, eds. H. Hirsch-Kreinsen, P. Ittermann and J. Niehaus, 35–51. Baden-Baden: Nomos.

Jia, Z., Y. Xiao, G. Shi, M. Wang, T. Lin and Shen, Z. 2019. Research on the Development Approach of Regional Manufacturing Industry in Internet+Era. IEEE Conference on Automation Science and Engineering, Vancouver, BC, Canada.

Kagermann, H. 2014. Chancen von Industrie 4.0 nutzen. In: Industrie 4.0 in Produktion, Automatisierung und Logistik, eds. T. Bauernhansl, M. ten Hompel and B. Vogel-Heuser, 603–614. Berlin.

Mertens, P. 2014. Industrie 4.0 = CIM 2.0? Industrie Management, 30(1): 27–30.

National Bureau of Statistics of China. 2019. Economic data series. Beijing.

National Science Foundation. 2014. Science and Engineering Indicators. Washington.

Nuernberg, J. and T. Wang. 2012. Implications of the 12th Five-Year-Plan for German Machinery Manufacturers. Frankfurt. Available at http://www.vdma.org/documents/105628/214296/12-07-11%20China_Studie_5_Jahres_Plan_Inhalt_K2.pdf.

Odenbach, J., E. Göll and S. Behrendt. 2017. Industrie 4.0 – Digital vernetzte dezentrale Produktion. Transformationsfeldanalyse im Rahmen des Projekts Evolution2Green. Berlin.

Orr, G. 2014. What Could Happen in China in 2015? New York, NY: McKinsey & Company.

Papula, J., L. Kohnová, Z. Papulová and M. Suchoba. 2019. Industry 4.0: Preparation of Slovak Companies. In: Smart Technology Trends in Industrial and Business Management, eds. D. Cagáňová, et al., 103–114. Cham: Springer.

Petermann, T., et al. 2011. Gefährdung und Verletzbarkeit moderner Gesellschaften am Beispiel eines großräumigen und langandauernden Ausfalls der Stromversorgung. Deutscher Bundestag, Ausschuss für Technikfolgen-Abschätzung, Drucksache 17/5672, Berlin.

Prause, M. and C. Günther. 2019. Technology Diffusion of Industry 4.0: An Agent-Based Approach. International Journal of Computational Economics and Econometrics, 9: 29–48.

Rauschnabel, P. A. and Y. K. Ro. 2016. Augmented Reality Smart Glasses – An Investigation of Technology Acceptance Drivers. International Journal of Technology Marketing, 11(2): 123–148.

Schlund, S., W. Mayrhofer and P. Rupprecht. 2018. Möglichkeiten der Gestaltung individualisierbarer Montagearbeitsplätze vor dem Hintergrund aktueller technologischer Entwicklungen. Zeitschrift für Arbeitswissenschaft, 72(4): 276–286.

Schneider, P. 2016. Welche Auswirkungen hat Industrie 4.0 auf KMU? Das Geschäftsmodell als Analyseinstrument. ZfKE – Zeitschrift für KMU und Entrepreneurship: Vol. 64, Geschäftsmodelle in kleinen und mittleren Unternehmen (KMU): 253–279.

Soder, J. 2016. Von CIM über Lean Production zu Industrie 4.0. In: Handbuch Industrie 4.0, Vol. I, eds. B. Vogel-Heuser, T. Bauernhansl and M. ten Hompel, 3–26. Berlin: Springer Vieweg.

Staufen. 2019. Deutsche Industrie 4.0 Index 2019. Köngen.

Terminio, R. 2015. Mr. Right's a Robot. HR in a New Era. German Chamber of Commerce in China. Beijing: German Chamber Ticker.

TÜV Rheinland. 2013. Szenarien und Kosten für eine kosteneffiziente flächendeckende Versorgung der bislang noch nicht mit mindestens 50 Mbit/s versorgten Regionen. Berlin.

World Bank. 2019. Economic data series.

Woetzel, J., G. Orr, A. Lau, Y. Chen, M. Chui, E. Chang and Qui, A. 2014. China's Digital Transformation – The Internet's Impact on Productivity and Growth. New York, NY: McKinsey.

Yudono, A., D. Satria and A. Erlando. 2019. Toward Inclusive Development through Smart Economy in Malang Regency. IOP Conference Series: Earth and Environmental Science, 328–335. Available at https://iopscience.iop.org/article/10.1088/1755-1315/328/1/012008.

chapter three

Smart manufacturing based on Digital Twin technologies

Shohin Aheleroff, Jan Polzer, Huiyue Huang, Zexuan Zhu, David Tomzik, Yuqian Lu, Yuan Lin, and Xun Xu
The University of Auckland, Auckland, New Zealand

Contents

3.1 Introduction

From Industry 1.0 to Industry 3.0, manufacturing has become more automated, computerized, and complicated (Kusiak, 2018). Industry 4.0, proposed by the German government in 2011 to promote the intelligence of manufacturing, aims at achieving a higher level of efficiency, productivity, and autonomy (Zhong, Xu, Klotz, & Newman, 2017). To achieve Industry 4.0, smart manufacturing was coined by several agencies, such as the Department of Energy (DoE) and the National Institute of Standards and Technology (NIST) in the United States (Thoben, Wiesner, & Wuest, 2017). According to NIST, smart manufacturing is "fully-integrated, collaborative manufacturing systems that respond in real-time to meet changing demands and conditions in the factory, in the supply network, and customer needs" (Thompson, 2014).

A key feature of smart manufacturing is the wide integration of product design, manufacturing, and general collaborative business across factories (Catarci et al., 2019). To realize the interoperability among these processes, every physical equipment in a factory needs a faithful virtual representation in the cyberspace. Physical entity and the virtual representation together with the connection form the cyber-physical system (CPS), which is one of the critical technologies of Smart Manufacturing, and the virtual representation is Digital Twin (DT) (Catarci et al., 2019; Lu, Liu, Wang, Huang, & Xu, 2020; Negri, Fumagalli, & Macchi, 2017).

DT in the cyberspace is not only an abstraction of the physical entity, reflecting its physical status, but also a brain of the physical entity. Combined with Big Data analytics and artificial technologies, DT achieves the ability of decision-making, which can help with the improvement of the performance of products or processes in the physical space (Qi & Tao, 2018; Tao, Zhang, Liu, & Nee, 2019). DT-based smart manufacturing has great benefits in several aspects (Tao, Zhang, & Nee, 2019):

- ***Reducing time to market:*** DT of products offers an efficient method to address the challenges of personalization.

- ***Increasing user engagement:*** DT as a middleware architecture allows interactive and information exchange between companies and consumers.
- ***Increasing visibility:*** DT provides high-fidelity models updated in real time. It allows operators to monitor the product and process in a clearer way with augmented reality (AR) and virtual reality (VR) technologies.
- ***Ensuring optimal operation:*** The ability of communication and decision-making of DT offers the possibility of optimizing operation processes in real time under different conditions.

This chapter elaborates DT's roles and opportunities in smart manufacturing via theoretical analysis and case studies. The remainder starts with an overview of DT, including its concept, research history, and current situation, followed by the researches and applications in manufacturing. Section 3 and Section 4 discuss how DT works at the product design stage and provides applications of DT for design and design verification. Section 5 and Section 6 present applications of DT for manufacturing equipment at the manufacturing stage. Section 7 discusses DT-based process automation and its case study on throughput optimization. Section 8 concludes the chapter and discusses the critical issues for future research.

3.2 *Digital Twin*

This section briefs the history of DT, clarifies its concept, and explains how DT can support smart manufacturing.

3.2.1 *Concept of Digital Twin*

DT emerged during digital transformation in the early 2000s and became popular for monitoring and predictions in manufacturing. It was proposed by National Aeronautics and Space Administration (NASA) in 2011 as "an integrated multi-physics, multi-scale, probabilistic simulation of a vehicle or system that uses the best available physical models, sensor updates, fleet history, etc., to mirror the life of its flying twin" (Shafto et al., 2012). The original intention of DT was to mirror the status of air vehicles with high-fidelity physical models, sensor data from the vehicle's onboard integrated vehicle health management (IVHM) system, maintenance history, and all available historical/fleet data obtained using data mining and text mining.

Inspired by the vision of DT from NASA, a significant amount of work has expanded the scope of DT in the manufacturing domain (Tao et al., 2019). Today's research intend to treat every digital version of a physical asset, equipment, or system as a DT. According to Madni (Madni, Madni,

Table 3.1 Four levels of Digital Twin

Level	Model description	Data acquisition and type
Pre-Digital Twin	Virtual system model focus on reducing technology/ technical-risk; physical twin does not exist	Not available
Digital Twin	A virtual model of the physical twin connects with physical twin	Performance, health status, maintenance; batch updates
Adaptive Digital Twin	A virtual system model of the physical twin with adaptive user interface (UI) and machine learning	Performance, health status, maintenance; real-time updates
Intelligent Digital Twin	A virtual system model of the physical twin with adaptive UI and reinforcement learning	Performance, health status, maintenance; both batch/ real-time updates

Source: Adapted from Madni et al. (2019).

& Lucero, 2019), four levels of the virtual representation were defined, as shown in Table 3.1. The Intelligent DT defined by Madni is the DT that has a high degree of autonomy, with the ability to analyze more granular performance, maintenance, and health data from the real-world counterpart.

Nowadays, DT refers to a virtual representation of manufacturing elements such as personnel, products, assets and process definitions, a living model that continuously updates and changes as the physical counterpart changes to represent status, working conditions, product geometries, and resource states in a synchronous manner (Lu et al., 2020).

Engineering research on DT has increased dramatically since 2016. The number of publications on this topic in 2018 tripled that in 2017 (Lu et al., 2020). Industry practitioners also showed great interest in DT. Microsoft brings out its DT products based on the Internet of Things (IoT) service platform (julieseto, n.d.). Siemens DT products focus more on product life cycle management (Digital Twin | Siemens, n.d.). Although there are many research works on DT academically and companies have launched corresponding products, the study of DT is still at an early stage. Current applications of DT mainly focus on process monitoring, history tracking, and simulation (Angrish, Starly, Lee, & Cohen, 2017; Boschert & Rosen, 2016; Lu et al., 2020; Revetria et al., 2019; Tan, Yang, Yoshida, & Takakuwa, 2019). Luo proposed a DT for a machine tool to improve its performance (Luo, Hu, Zhang, & Wei, 2019). A DT-based solution for tracking damage of military vehicles was implemented based on a health tracking system (Mars, Suter, & Bauman, 2018). In the context of future factory, DT offered a method of exploiting the full potential of modeling and simulation within smart factories (Weyer, Meyer, Ohmer, Gorecky, & Zühlke, 2016).

3.2.2 Reference model

DT reflects the two-way dynamic mapping between a physical entity and its digital model in cyberspace (Angrish et al., 2017). As a middleware architecture, the essence of DT is to abstract the physical entity into the cyberspace to provide basic information for the high-level management system to make real-time decisions for production (Lu & Xu, 2018). To provide a sufficient synchronization mechanism between digital models and physical objects for implementation of DT, a reference model consisting of three components was proposed in 2018 (Lu & Xu, 2018), as shown in Figure 3.1.

- ***Information model:*** The physical entity is abstracted according to the predefined information model that represents its specifications of concern. The standard for product data representation and production data exchange plays an important role.
- ***Communication mechanism:*** A synchronized bidirection data transfer mechanism is a basic requirement for developing a DT. Considering the reaction time needed for different levels of

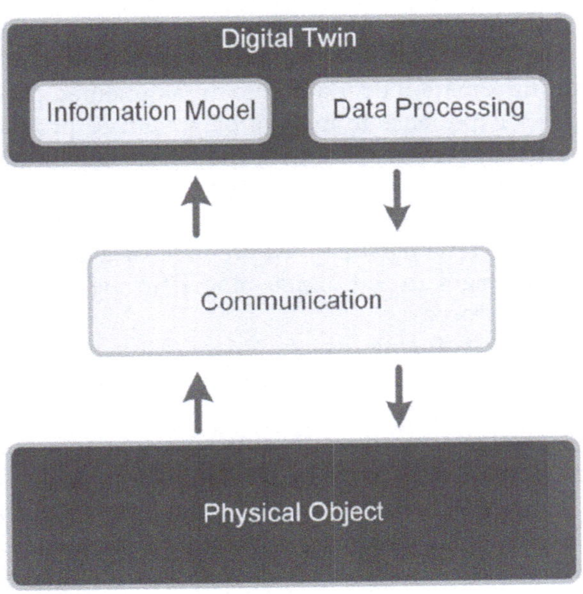

Figure 3.1 Reference model of DT. (Adapted from Lu & Xu, 2018.)

communication, three classes of real time were defined (Neumann, 2007). Soft real time with scalable cycle time is used on the shop floor for monitoring and process automation; hard real time, usually with cycle time of 1–10 ms, is used in control applications; isochronous hard real time, with cycle time 250 μs–1 ms and jitter less than 1 μs, is usually implemented in motion control.

- ***Data processing module:*** Large volume of data gathering from the physical world, also called Big Data, needs certain processing methods to deal with (Luo et al., 2019). A low-latency data processing system that can integrate domain knowledge verification for data processing is necessary for developing a DT.

3.2.3 *Digital Twin for smart manufacturing*

DT now has far more significance from production design to product services than its original version proposed by NASA. It is a key enabling technology to satisfy the design, manufacturing, and field services of highly personalized products. Regarding the vision of smart manufacturing, a DT has an impact on smart manufacturing in the following aspects.

- ***DT for design:*** With Big Data obtained in the entire product life cycle and real-time connectivity, DT lands a role in product design, facing personalized manufacturing strategy. The data collection and interaction with customers for personalization in virtual space being driven by IoT and smart systems. Providing transparent information by useful information overlaid onto an augmented product gives a unique user experience to make a personalized product more feasible ever in the era of the fourth industrial revolution. DT brings extra value beyond virtual prototyping by using a Computer-aided Design (CAD) model by closing the gap between customers who are looking for physical personalized products and manufacturers who can make changes in the digital model through an agile product development process.
- ***DT for manufacturing assets:*** Manufacturing assets, such as machine tools, robots, and conveyors can be connected to their DTs in cyberspace via networks. A clear observation of real-world performance is available and operating status can be monitored through real-time data capture from the physical world. Based on the information gathered from the physical entities, situational awareness can be improved, the physical assets can provide reliable predictions of health and safety conditions, operation resilience and flexibility can be enhanced, and humans can seamlessly and comprehensively interact with both physical assets and its DT.

- ***DT for manufacturing process:*** DT of physical entities on the shop floor allows the establishment of a self-organizing environment with connectivity and data tracking abilities.

Sections 3 through 7 elaborate the role of DT for smart manufacturing and possible application scenarios.

3.3 Digital Twin-enabled product personalization

In the era where cutting-edge technologies are a natural part of the service experience, personalization will change depending on the preferences of individual customers. Companies are adopting innovative digital technologies, such as DT interactions through IoT (Bolton et al., 2018). Digital technologies can provide a highly personalized and immersive environment that allows interactive and rich information exchange between organizations and consumers (Parise, Guinan, & Kafka, 2016). Definitions of DT technology underlined two characteristics. First, the connection between the physical and virtual space. Second, real-time data connection by using sensors.

3.3.1 Digital Twin-based agile product development

Industry 4.0 enables the entire CPSs where new products are digitally developed by creating a virtual copy of the physical world. Over the IoT, CPSs interact in real time to a responsive and customer-driven approach. Therefore, the impact of data-driven manufacturing gives the impression of iterative stages. Figure 3.2 shows an iterative incremental product development using DT. In this methodology, IoT, Big Data, and Artificial Intelligence (AI) technologies empower customers to create a cloud-based customer profile (CCP) by capturing preferences in the digital space through real-time interactions with the physical world. The result represents a live list of functionalities and appearance called customer backlog, which reflects all the desires in a digital space.

Using DT not only saves resources (time, cost, materials) but also significantly increases interaction with customers through data-oriented user experience. For instance, by using DT, a customer has the opportunity to choose different parts such as saddle, drop handlebar, frame, and pedal to make a personalized bike. Therefore, a customer can apply iterative incremental changes in shape and color of a single or joint part of a bike to create an augmented experience and make a decision. Utilizing Industry 4.0 to build DT brings a tangible value to agile product development. Figure 3.2 shows that DT involved in consequence of (1) DT analysis, (2) DT test, (3) DT design, and (4) DT build resulted in

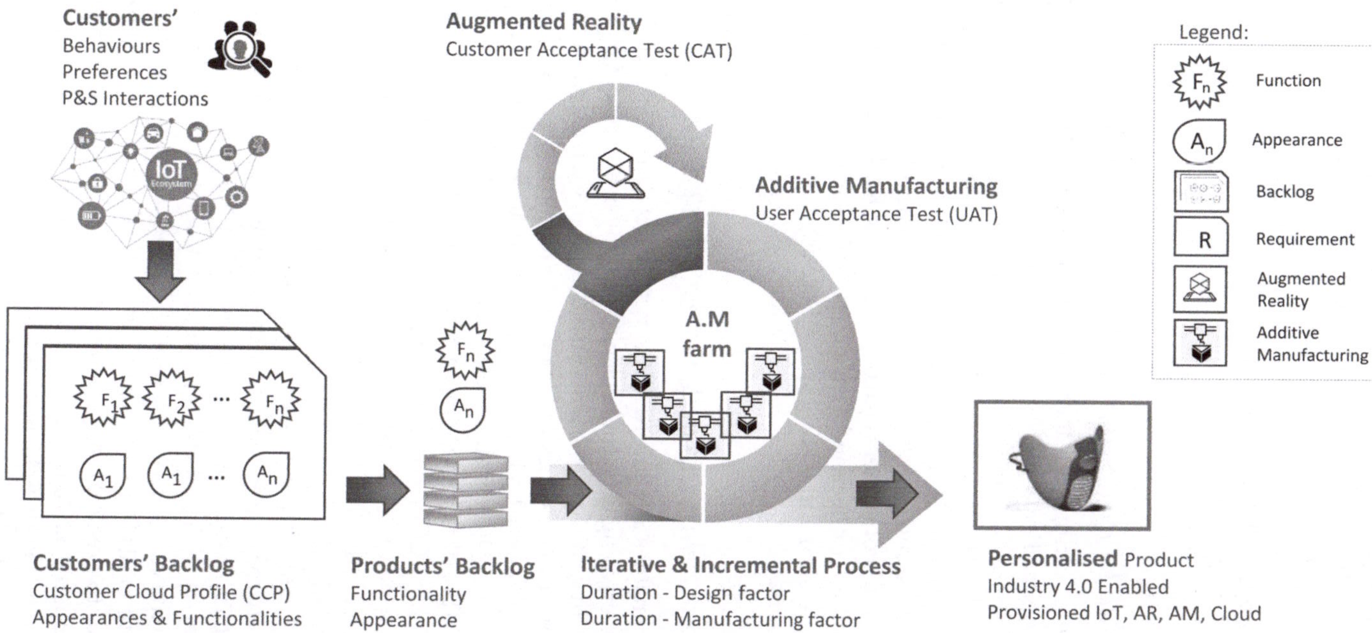

Figure 3.2 Industry 4.0 Agile Mass Personalization Methodology (AMPM).

agile product development to achieve personalization under Industry 4.0. DT-enabled product development empowers customers to visualize final product, create product changes in the augmented space, and make decisions based on real-time information about the consequence of changes to a product.

The entire agile product development is empowered by IoT to provide real-time interaction for further personalization. Table 3.2 shows the core Industry 4.0 technologies for implementing DT to adopt agile product development.

3.3.2 *Digital Twin-enabled product personalization*

Industry 4.0 embraces data exchange in advanced manufacturing by utilizing technologies such as Cloud, IoT, AR, Big Data, AI, AM, and Autonomous Robots. Due to the nature of customer-oriented product development, keeping state information during personalize product development required technologies to manage attributes; e.g., metadata, configurations, and conditions. Consequently, DT creates and keep changes for both customers and providers at the same time. A DT is a digital replica of desired product/service, which gradually succeeds customer satisfaction in virtual space through iterative incremental product development life cycle. Unlike most definitions, DT holds wanted attributes; e.g., functionalities and appearances of a wanted product/service. A JavaScript Object Notation (JSON) file can keep data to describe DT at any time. The following characteristics are associated with DT:

- Tracking of product/service and environmental conditions.
- Adopted iterative and incremental process.

***Table* 3.2** Industry 4.0 technologies-enabled DT for product personalization

No.	Process	Technology	Result
1	Data Collection	Internet of Things	Customer Cloud Profile
2	Product Backlog	Artificial Intelligence	Provide trade-off proposal
3	Customer Backlog	Cloud	Update repository of needs
4	Iterative & Increment	Augmented Reality	Visualize parts in each cycle
5	Manufacturing	Additive Manufacturing (AM)	Make a personalized product
6	Performance	Edge Computing	Monitor performance

- Its seamless link with the corresponding wanted product/service.
- Keeping CCP live and up to date.
- Provision and development by using IoT, Big Data, and AI technologies.

To make Physical Twin (PT) feasible, core Industry 4.0 is required. The role of IoT is crucial to provide sufficient data about desire functionalities and appearances by synchronizing physical things and virtual representatives in cyberspace. Therefore, DT connects the cyber and physical world instantly by collecting real-time data using Edge computing and embedded sensors. The PT can represent how physical things operate in cyberspace even though this is not a clone version. DT is not just a CAD or a digital model. It is a virtual representation of how necessary attributes respond to changes.

Personalization is highly dependent on user/customer experience (UX/CX) based on tailored functionalities and appearances. The core Industry 4.0 technologies including but not limited to IoT, Big Data, AR, AM, and DT make suitable personalization. The former used for manufacturing of personalized products and the latter used during/after product development in the UX/CX process.

3.3.3 *Case study*

Personalized product development as an emerging manufacturing paradigm gives customers the freedom to take part in the product development process (Zheng, Yu, Wang, Zhong, & Xu, 2017). Historically, personalization has meant that the product delivery time and cost are significantly high. To tackle product personalization challenges, an experiment was undertaken using a DT-enabled agile product development methodology. This study focuses on air pollution facial mask development, which is ideally developed based on individual face profile. Figure 3.3 shows DT implementation by using ThingWorx as an IoT platform, Creo for designing a CAD file, Cloud for keeping CCP updated, Unity as an Augmented Reality platform, 3D printer for Additive Manufacturing (AM), and Edge computing form enabling Big Data analytics.

A DT is built by seamless integration of IoT and AR through Agile Mass Personalization Methodology (AMPM). ThingWorx is an industry IoT platform and offers a range of Software Development Kits (SDKs) and Application Programming Interfaces (APIs) for integration to other systems. A sensor is embedded in the facial mask and connected to the ThingWorx by using the Android SDK, which is shared by the sensor's supplier. Figure 3.4 displays how ThingWorx as an IoT platform is

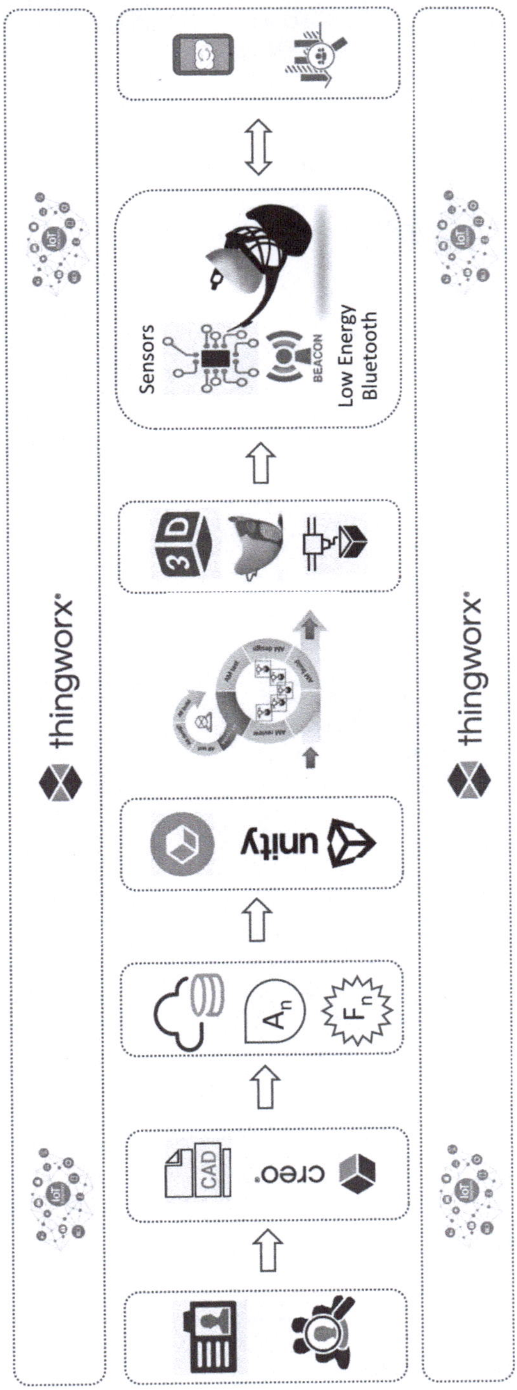

Figure 3.3 DT-enabled product personalization under Industry 4.0.

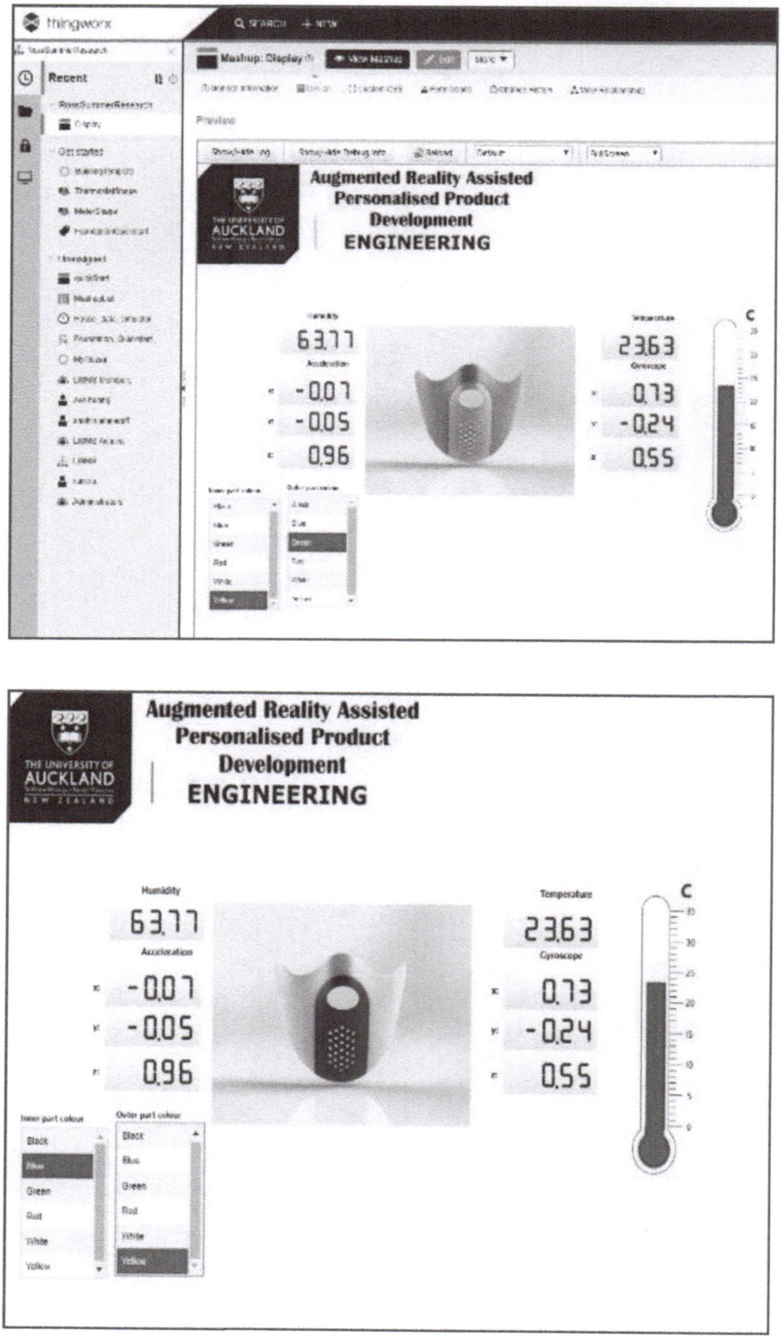

Figure 3.4 DT-enabled personalized facial mask.

utilized for making changes and providing real-time information. Then customers' changes are reflected on a web-based and mobile application. Finally, an augmented experience for personalized (Different Inner and Outer colors) facial masks was provided and the real-time information overlaid on the screen. In the end, the customer has access to both real-time and historical data for further Big Data analysis and added value services.

Both Unity and OpenCV are used to develop the AR application for visualizing a tailored facial mask. An open-source library was used in Unity for facial tracking to make a real-time augmented experience after a change (color as an indication) by a customer. To model a real-life facial mask, we connected a thing as a DT to the ThingWorx server through mobile using an edge device. Then a Thing Shape called Mask is created like a template for creating new (personalized) masks. The mask shape has the properties of Color (number), Color Scheme (text), List of Colors (custom), and Size (number). The Color is a numerical representation of the color of the mask, the Color Scheme is the text description of the color of the mask, and the Size is a number representing the sizing of the mask. Figure 3.5 displays a list of colors that consists of two smaller data shapes layered on top of one another.

The outer of the smaller data shapes called MaskColors represents the inner colors, and the inner of the two data shapes called OuterColors represents the outer colors. OuterColors contains a media entity, which is used to store an image of the color scheme on a mask. Figure 3.6 displays a web-based interface that allows a customer to see the real-time data of the DT (model of the facial mask) from the sensor.

To enable a customer to visualize iterative and incremental changes within the agile product development process, the real-time data from the sensor is overlaid on a personalized facial mask by passing the data to the Unity as an AR application. By using OpenCV and webcam, the Unity application can track the facial mask and customer's face and eyes, and then overlay the AR aspects in the correct place.

3.4 *Digital Twin-based design validation*

Product development is an iterative process involving design, fabrication, and evaluation. In a traditional workflow, there is no strong connection between each development stage. Therefore, only a limited genre of design issues can be revealed at each stage. For instance, it is challenging to design with complete fabrication concerns or quickly adjust fabrication settings without defying design purposes. A complete product design validation must be carried out before hitting production. Such inability to make corrections in time often leads to long development cycles and a significant waste of resources.

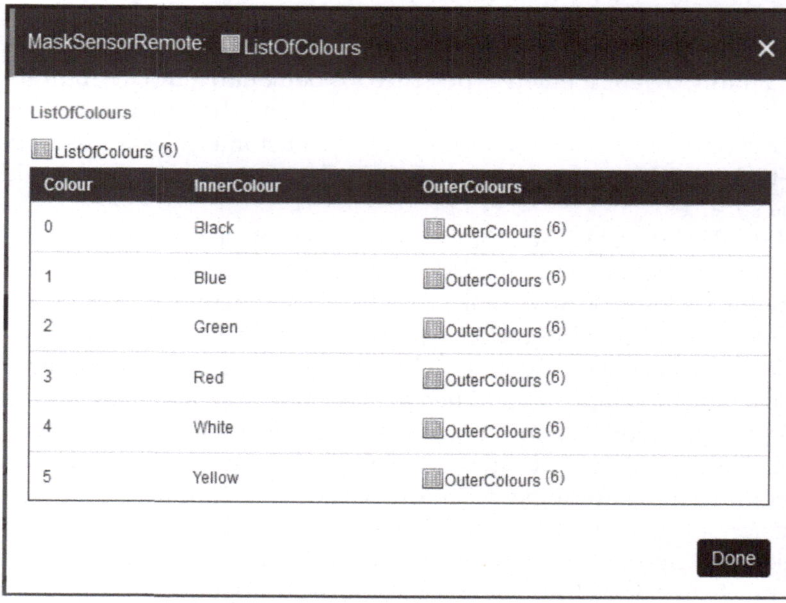

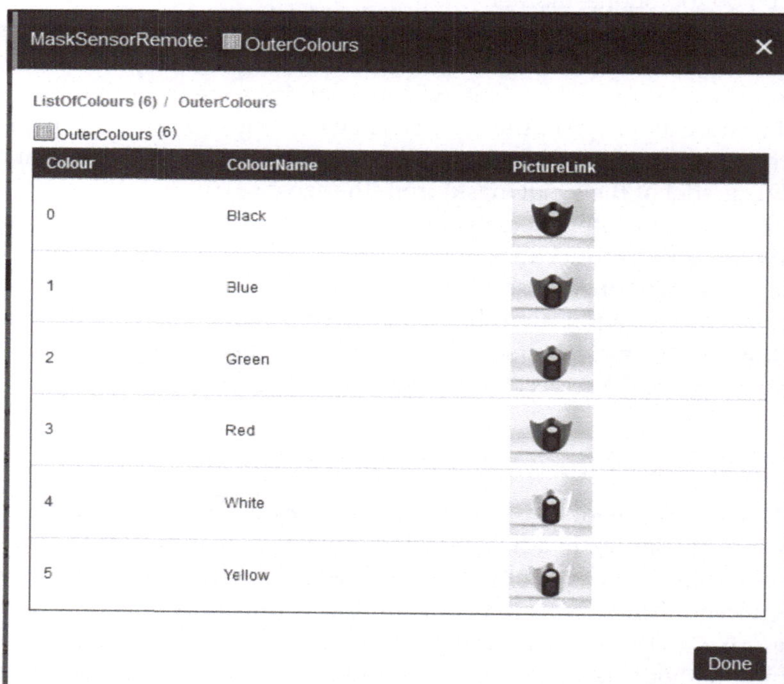

Figure 3.5 (Top) Inner colors data structure. (Bottom) Outer colors data structure.

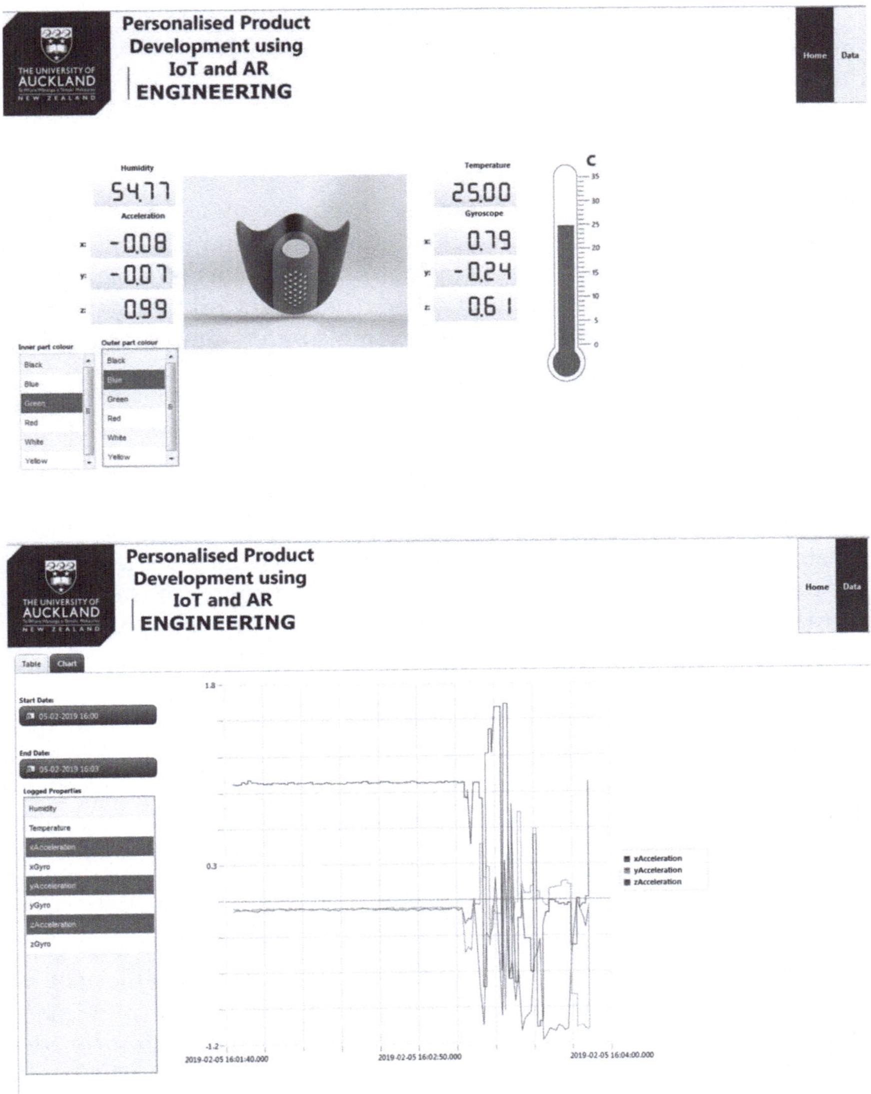

Figure 3.6 (Top) Real-time data of user interface. (Bottom) Historical data.

The utilization of DT can facilitate more effective information flow, and therefore will be able to help the industry tackle these barriers. Specifically, by having a DT of product design, operators responsible for other product development stages can get involved and provide feedback more efficiently. Meanwhile, a digitized development stage is highly configurable so that it can be quickly adapted accordingly.

This section will begin with a brief review of related work, then it will propose a framework for DT-based design validation. A case study will be discussed at the end.

3.4.1 *Related work*

The DT-based design validation is made possible by some emerging technologies such as VR, AR, AM, AI, cloud computing, and IoT. Researchers have been implementing these technologies for different tasks. Significant improvements in concurrency and coherency of the workflow have been achieved in different studies. Some examples are summarized below according to the design phases they are focused on.

3.4.1.1 *In the early design phase*

Fabrication of physical prototypes is an essential process of design validation. However, Physical Prototyping (PP) is often costly and time consuming, sometimes even impossible. Therefore, researchers exploited VR to construct digital environments for engineers and end users to validate key design concepts (Berg & Vance, 2017). Such an approach is known as Virtual Prototyping (VP) (Wang, 2002), which is particularly useful for high-value products such as automobile vehicles (Kim, Kashiba, Dai, & Shiraishi, 2017).

Modeling techniques are the foundation of conventional engineering simulation, which, according to Wang (2002), is a part of VP. Although more comprehensive modeling should lead to better optimization results (Pellicciari, Berselli, & Balugani, 2015), the extra computation time may defy the purpose of improving efficiency (Evans, Castellazzi, & Johnson, 2016). Therefore, coarse modeling methods (Vergnano, Berselli, & Pellicciari, 2017) were introduced for VP.

Besides the direct collection of feedback using VP, the indirect involvement of engineers and end users can further improve the efficiency of design. Specifically, by having them constructing DTs that describe manufacturing capabilities and end user demands, designers can have adequate guidelines for generating a more mature initial design, which will eventually reduce the number of design iterations.

To avoid the extra delay caused by outsourcing fabrication tasks, AM is often adopted for producing prototypes in-house (Lanzotti, Carbone, Grazioso, Renno, & Staiano, 2018). A Design for Additive Manufacturing (DfAM) Worksheet (Booth et al., 2017) was proposed to provide aid for designers. A DfAM ontology (Kim, Rosen, Witherell, & Ko, 2019) has been developed as a knowledge base for providing design guidelines using Machine Learning.

The manufacturers today are required to cope with more versatile customer demands. Having DTs of customers is the prerequisite for automatically adjusting and validating the design for each individual. The section "DT-enabled product personalization" provides more in-depth discussions about this.

3.4.1.2 During fabrication

Even if the designer has strictly followed the design guidelines, some issues may still be inevitable during fabrication. For instance, the inconsistency of raw material or environmental change may influence the yield rate. The in-process defect detection is particularly crucial for AM due to its layer-based nature.

A DT is the foundation for automatic defect detection because many issues can only be revealed by comparing the actual workpiece against the DTs of product design. An online defect detection method was introduced for metal 3D printing (Fang et al., 1998), which was based on 2D imaging. Cheng and Jafari (2008) and Bukkapatnam and Clark (2007) proposed online process control methods using 3D imaging. The extrusion process was modelled in the work by Holzmond and Li (2017). To avoid the needs of complex models, AI was adopted for problem identification (Delli & Chang, 2018; Jin, Zhang, & Gu, 2019).

With highly automated facilities, manual operations are still necessary when a non-recoverable defect is detected. In this scenario, a more effective human-machine interface (HMI) can reduce machine downtime. AR can improve HMI in two ways, namely enhancing human perception by showing detailed insights as information overlays and accelerating the execution of human decisions by supporting intuitive control inputs. MIT Media Lab developed an AR-enabled user interface (UI) for a 3D printer (Eiriksson et al., 2017), which could use design information to generate an AR preview to compare with actual outcomes. They also leveraged fiducial markers to provide intuitive UI on tablets. Similar marker-based AR techniques have been used in the work by Ceruti, Liverani, and Bombardi (2017) to help the operator making decisions.

3.4.1.3 Under evaluation

The primary purpose of design evaluation is not only comparing prototypes against design specifications, but also exploring different design options (Camburn et al., 2017).

The initial design has a decisive influence on the development of a sophisticated system. With DT, it is possible to evaluate Mixed Prototypes (MP) (Bordegoni, Cugini, Caruso, & Polistina, 2009), which is a mix of physical and virtual components. Without reliance on an entirely fabricated prototype, the evaluation can be started earlier in the development process. When the tester only needs to focus on the fabricated key functional components, the troubleshooting process is more straightforward (Buchholz, Vorsatz, Kind, & Stark, 2017).

Aside from performance, the user experience would also affect the usability of the final product. Since DT contains a well-defined correlation between each component, unfabricated parts can be visualized through Mixed Reality, so the nonfunctional aspects like appearance and UI can be evaluated on a mixed prototype as well. The virtual parts of MP can be updated without much additional cost, so the designer can quickly explore different design options and easily tailor the design for specific user groups when necessary.

3.4.2 *Proposed framework*

Here we propose a framework to demonstrate the key values of utilizing DTs in design validation. Two types of DTs are presented in this framework. First, the DT of product design itself is the essence, which is developed based on CAD models with extra information. Second, each development stage will have its own DT, or at least include some elements that have DTs. For instance, the equipment in the fabrication stage will have its own DT.

The main design principle is to make the system human-centric and distributed. Being human-centric requires the system to provide an intuitive and efficient UI. AR was utilized to achieve this purpose. To support distributed collaborative product development, the system should be cloud-based. The cloud platform enables different collaborators to exchange information smoothly.

As shown in Figure 3.7, the overall system comprises three modules covering design, fabrication, and evaluation. A cloud platform initiates the communication between every module. Each module has a dedicated AR-enabled HMI for the user to operate with DTs.

- ***Design module:*** The design module has three major functions. First, with dedicated adapters to interface with different CAD systems, the design module enables the designer to embed extra design information into a DT. Second, it presents information gathered from

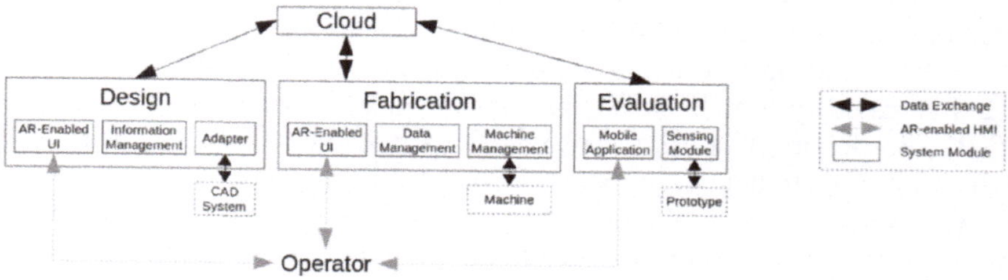

Figure 3.7 System overview.

fabrication and evaluation stage as the design guidelines. Third, it supports intuitive manipulation of CAD models through AR.

- ***Fabrication module:*** The key to the utilization of DTs in fabrication is to construct DTs for the facilities. The fabrication module is a containerized software package that can be easily deployed onto different equipment. The whole package consists of three parts. The first part is connected with a specific machine and provides intelligence for problem handling. The second part is a local database responsible for logging the status of both the machine and the workpiece. The third part is a lightweight web server hosting a web-based AR interface to communicate with the operator.
- ***Evaluation module:*** The evaluation module has two components, namely, a configurable sensing module and a mobile application. The former is responsible for gathering data from the prototype, whereas the latter connects the evaluation module with other modules.

3.4.3 Case study

The proposed framework has been validated in a detailed implementation. Specifically, the construction and application of a DT for both product design and development stage have been fulfilled.

As demonstrated in Figure 3.8, we developed a stand-alone plug-in for SolidWorks to function as the design module. This module assists

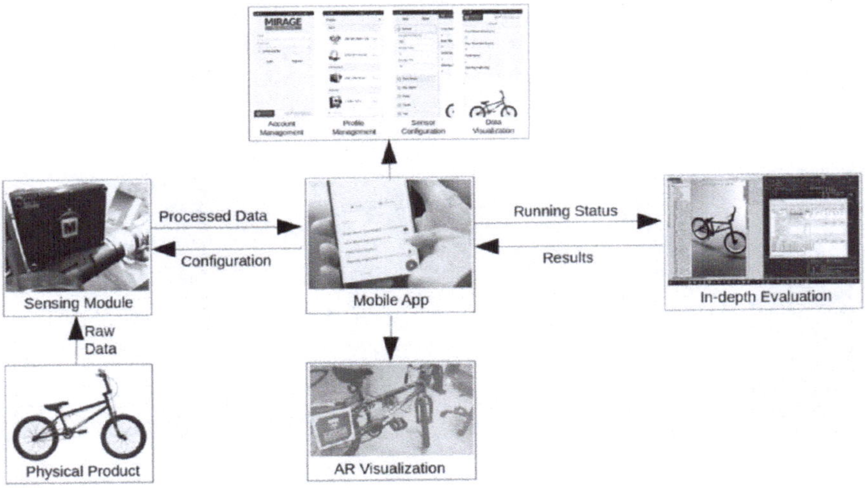

Figure 3.8 The utilization of a DT in design and evaluation stage.

the designer in constructing a DT of the product design. It automatically parses the loaded CAD assembly file, and then lists all detected moveable parts and available sensors. Therefore, the designer can easily define a series of mapping from each part to the sensor responsible for measuring its movements. This information is stored as a product profile for interpreting sensor readings during evaluation.

A configurable sensing module based on ESP32 and a mobile application for Android OS were developed as the evaluation module for helping a tester to evaluate the prototype. After login into the cloud platform, the tester can manage a list of product profiles that represent the prototypes to be tested. The tester can adjust the configuration of the sensing module and check the real-time sensor readings either in plain text or AR.

When any unexpected reading is detected, the tester can use the application to forward the running status to the designer. Then, the designer could use the design module to drive the CAD model with actual sensor data for in-depth evaluations. Once the analysis is finished, the designer can send the results and additional instructions to the tester for adjusting test settings or properly reporting the issue.

As shown in Figure 3.9, we constructed a DT for the fabrication equipment as the fabrication module. The database is a crucial component to keep data for both machine management and user interaction.

In the beginning, the design model was converted into the initial sequence of machine commands according to design requirements and machine capabilities. During the machining process, by evaluating the actual status of the workpiece and the machine, the DT of the machine can make judgments about the current situation and adjust the following machine commands accordingly.

When a non-recoverable problem is detected, the machine will abort the current job and raise the issue. The AR-enabled UI helps the

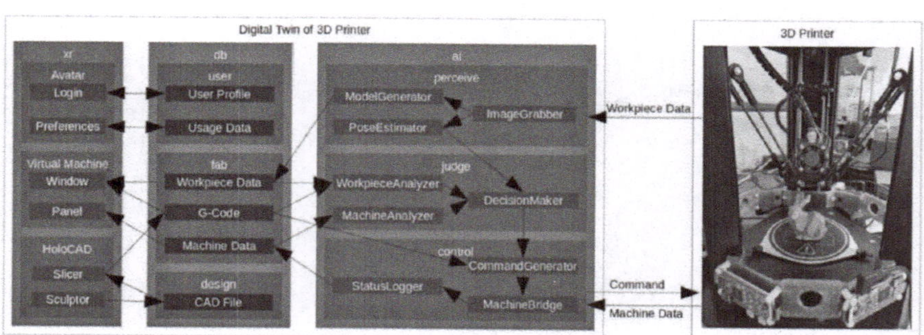

Figure 3.9 The utilization of DT at the fabrication stage.

machine operator to understand the cause of the issue. Then, the operator could either reinitiate the fabrication process with adjusted settings or abort the task and forward the issue to the designer regarding fabrication feasibility.

3.5 Digital Twin-based machine tools

Industry 4.0 is considered to be the fourth industrial revolution, and its main focus is based on smart manufacturing (Zhou, Liu, & Zhou, 2015). The term originated from a strategic initiative report of the German government in 2012 (Kagermann, Helbig, Hellinger, & Wahlster, 2013, p. 5). Figure 3.10 shows the evolution history of industry since the 18th century; to compare with traditional manufacturing industry, Industry 4.0 comes with more requirements for the industry, such as shorter development periods, individualization on demands, flexibility of production line, and decentralization and resource efficiency (Lasi, Fettke, Kemper, Feld, & Hoffmann, 2014).

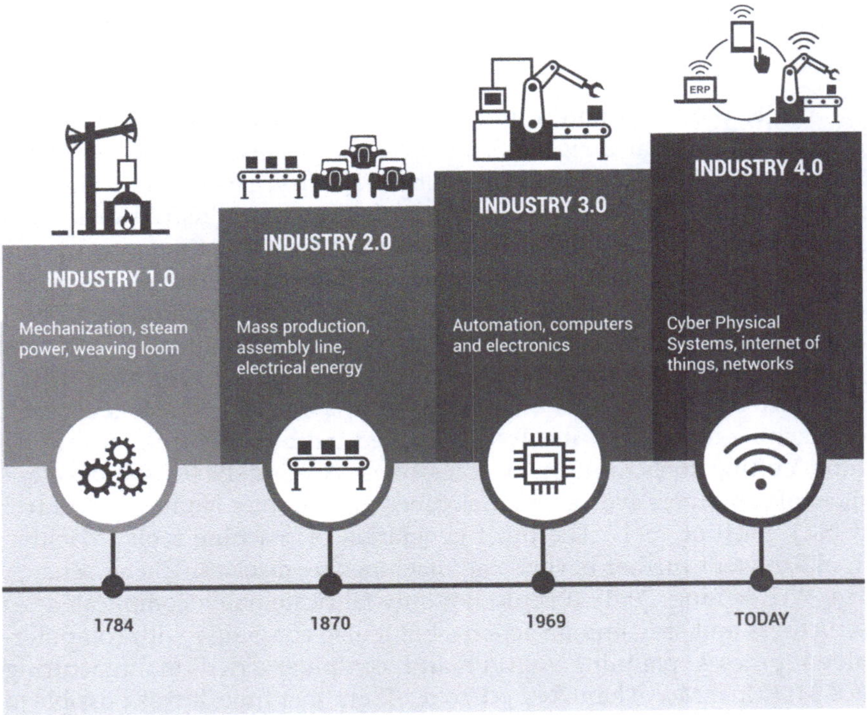

Figure 3.10 From Industry 1.0 to Industry 4.0. (Adapted from Lasi et al., 2014.)

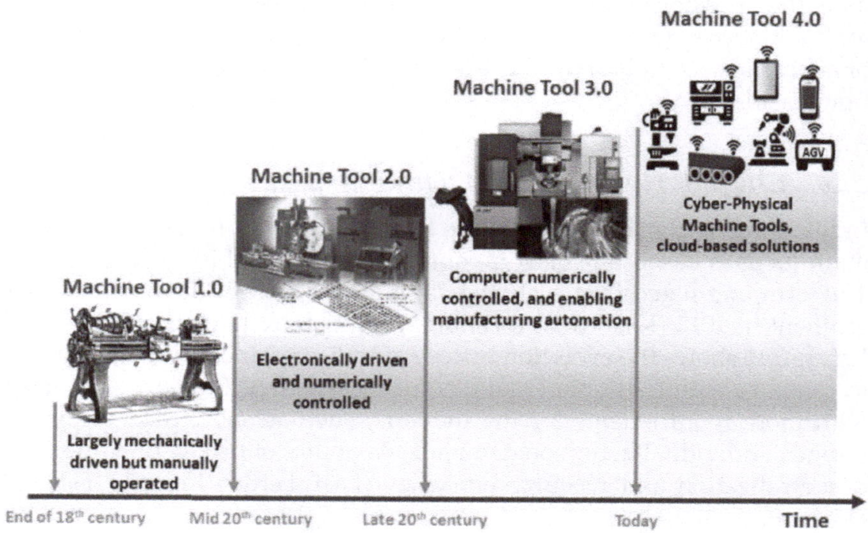

Figure 3.11 Machine Tool 1.0to Machine Tool 4.0: Evolution of machine tools. (Adapted from Xu, 2017.)

Under the revolution in industry, the evolution of machine tools is also moving forward. Figure 3.11 shows the evolution of machine tools accompanied by the industrial revolution. The first generation of machine tools (Machine Tool 1.0) was linked with the first industrial revolution from 1700s to the early 1800s. The machine tools at that time were mostly manually operated and required some essential skills and experience to control them. In the early 1950s, machine tools had experienced significant advancement in motion actuation and control. The manual control was gradually replaced by numerical control (NC), and the electronic actuation started to be deployed on most machine tools. The second generation of machine tools (Machine Tool 2.0) saved more manpower, machining time, cutting tools, and other accessories. Around 1980s, the development of computers evolved NC machines to Computer Numerical Control (CNC) machine tools. The third generation of machine tools (Machine Tool 3.0) went further beyond the machines themselves (Cheng, Zhang, Hu, Wu, & Yang, 2001). It could not only fabricate much complicated 3D structures and save human actions, but it also combined with computer-aided process planning (CAPP) and computer-aided manufacturing (CAM) (MuchMoreThanDNC.pdf, n.d.). From that time, it was possible to integrate design and manufacturing (Xu, 2008).

Throughout the evolution process of machine tools from Machine Tool 1.0 to Machine Tool 3.0, efforts to make machine tools faster, more accurate, reliable, flexible, and safer have never stopped. Nevertheless, as we step into the era of Industry 4.0, an urgent need to advance current CNC machine tools to a higher level of connectivity, intelligence, and autonomy has also been raised. Triggered by the recent advancements of CPS, IoT, and Big Data analytics, Machine Tool 4.0 will be defined as smarter, well connected, widely accessible, more adaptive, and more autonomous. "With MT 4.0, total vertical and horizontal integrations become possible. Machine tools will no longer exist as a piece of isolated manufacturing equipment; they can be service and solution providers" (Liu & Xu, 2017, p. 72).

3.5.1 Digital Twin-based machine tool

DT has been applied in many industrial fields since the concept was proposed, and it has shown its enormous potential in the manufacturing environment. In the early stage, the application of DT mainly focused on the industry of aerospace. The US Air Force Research Laboratory applied DT to create a high-fidelity flight model, by integrating data from both physical and virtual parts to generate a more accurate prediction of fatigue (Bielefeldt, Hochhalter, & Hartl, 2015). DT has also been applied to generate the model of aircraft wings to detect and monitor the damages and defects in aircraft structure. The results have been approved to have higher efficiency and accuracy than the traditional method (Tuegel, 2012).

Furthermore, Tuegel et al. (Tuegel, Ingraffea, Eason, & Spottswood, 2011) proposed the Airframe Digital Twin (ADT) concept to reduce the unnecessary costs during the aircraft's maintenance. Apart from aerospace industry, DT has been involved in so many various stages in manufacturing, such as product design, process planning, machining process, assembly process, and assessment process. Recently, Liu et al. (Liu, Vengayil, Zhong, & Xu, 2018; Liu, Xu, Peng, & Zhou, 2018) implemented the concept of DT into the machine tool; DT can take full advantages of the real-time data collected from the physical world and allows the physical machine tool to get the intelligent and autonomous functionalities.

Compared with traditional machine tools, DT-based machine tools allow effective information collection and data exchange from its physical part and its DT, it also enables the intelligent Big Data analytics and high-level interaction between machine tools and operators. Therefore, DT-based machine tool should have the following features:

- ***Well networked and distributed in the system:*** The connection and distribution is limited not only between the physical part of machine tools but also between their twins and, most importantly, the connection between the physical part and its DT. Both physical machine tool and its DT could utilize the information and data from different machine tools and DT, to build up a holonomic communication network.

- ***Adaptive and predictive during the machining process:*** Through data analytics and intelligent algorithm, collected information and data will be allocated for different using purposes. Depending on the historical data and real-time data from DT-based machine tools, the machine itself is looking for the prediction of health and safety conditions in the future, to achieve the self-correction and evolution.
- ***Humans in the loop:*** Humans should be able to interact with both physical machine tools and their DTs. Machine tools will distribute different types of information and different people with different roles. The information can help humans to more effectively operate, service, and manage machine tools.

In the rest of this section, we will discuss the architecture of DT-based machine tool, and how does the machine tool utilize the DT technology. A case study is also included proving the feasibility of the concept.

3.5.1.1 Through data communication and management

Aiming to develop DTs of machine tools for smart manufacturing, Machine Tool 4.0 and DT requires interoperable data communication, not only between the machine tools but also between machine tools and other system and operator, the desired DT should also endow the machine tools with self-awareness, self-prediction, and self-comparison capabilities. To achieve the characteristics of DT, the DT data needs to be more comprehensible, reliable, and intelligent than traditional manufacturing data. The DT data can usually be divided into three categories: static property data, real-time data and measurement data (Augment, 2016). Static property data stand for the basic properties of the physical part, such as the information of machines, cutting tools, workpieces, and the physical environment. This part of the data can represent the physical part of DT. For real-time data, it can represent the status of the operation process from different aspects. The physical part can usually provide some real-time data during the operation. However, if we want to achieve ultrahigh fidelity to reflect the physical part, the DT data requires a huge number of data acquisition devices (DAQs) and sensors to retrieve real-time data from the physical world. The retrieved real-time data and historical data can represent the virtual part of DT. Measurement data refers to the measurement results obtained from different measurement devices during the operation (Augment, 2016). The measurement data can be used to monitor and optimize the operation process, it will provide useful information to DT to control both the physical part and virtual part of it.

3.5.1.2 *Interaction between a physical machine tool and its Digital Twin*

There are three critical components for the DT-based machine tools, as shown in Figure 3.12. First is the physical part, which is the machine tool itself, and machining processes. The second part of the system is obviously the DT of the physical machine tools. Based on the physical machine tool, we can directly collect much information and data. However, to have more precise and realistic DT of the physical machine tool, extra DAQs are required to acquire more specific data and information before, during, and after the machining processes. All the real-time data acquired from the physical machine tool and attached DAQs will be part of its DT. However, the meaning of DT should be more than this, it will collect and store all the real-time data, doing further analysis based on these data and historical data to make predictions on future performance of the physical machine tool. The last part of the system will be the intersection and convergence between physical machine tools and its DT. Both physical and virtual parts exist in DT, the virtual part can keep ultrahigh synchronization and fidelity to reflect the physical part. The intersection and convergence also happen between real-time data and historical data, which can make the DT data more comprehensible and useful. In addition, DT can collect and update data in real time, the virtual part can continually self-improve by comparing virtual part with a physical part in parallel. The future prediction and comparison result is sent back to the physical machine tool or the operator to achieve the self-correction and self-evolution process.

3.5.1.3 *Augmented reality-assisted visualization*

Depending on the descriptions of DT and the DT data, it is sometimes difficult for people and operators to easily understand, interact and make decisions based on this huge amount of data. Therefore, the visual representation is important for them to access the DT data, to interact with

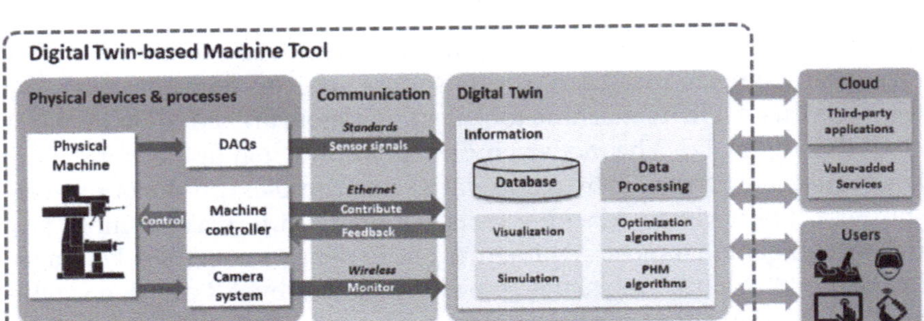

Figure 3.12 The architecture of DT-based machine tool.

Figure 3.13 The application of Augmented Reality in smart manufacturing. (Adapted from Hoover, 2018; Makris, Karagiannis, Koukas, & Matthaiakis, 2016.)

both physical machine tools and its DT and make better decisions during the operation process. Furthermore, visual representation can represent DT and the DT data in more comprehensible and intuitive way. AR as one of the key technologies in Industrial 4.0; it allows the user to see the physical world, with the virtual objects composited with the physical world. It includes multiple sensory modalities, such as visual, auditory, haptic, somatosensory, and olfactory. AR is currently widely implemented in the manufacturing field, and manufacturing is one of the fields where AR can be used to improve the current techniques and provide solutions in the future. Figure 3.13 below shows some applications of AR during machining process and engineering design. With the rapid development of computer and manufacturing technologies, to allow users and operators to visualize and interact with the manufacturing information associated with the manufacturing processes becomes a trend. Therefore, due to the characteristics of AR technology, it can seamlessly integrate with DT, it becomes a critical bond between physical and virtual part, to flawlessly achieve the interaction and convergence between machine tools and its DT.

3.5.2 *Case study (AR-assisted intelligent machine window)*

In the era of smart manufacturing, machine tools become part of a platform that connects machines, systems, data sources, and operators. All the decisions made by machine tools and operators are based on Big Data analytics and its DT. The modern industrial production in the world market requires a further increase in variety of complex products in ever-shorter delivery times. Machine tools are required to become more and more intelligent in handling this situation. To demonstrate the advantages of DT based machine tools, the case study named by AR-assisted intelligent machine window has been created.

In this case study, an AR-assisted Intelligent Window was developed on a touch-screen computer for an EMCO Concept 105 Milling

Figure 3.14 AR-assisted intelligent machine window setup.

Machine, CNC controller of this machine tool is a Fanuc 21 CNC which has built-in sensors that can report portion of real-time machining parameters (e.g. axis positions, spindle speed) and error messages. The whole setup is shown in Figure 3.14. The CNC machine is connected to a host PC through Ethernet cable. The host PC is also hosting a server that stores the real-time and historical data. The Server on the host PC can retrieve the real-time working data from the CNC machine and the extra data from external acquisition devices and multiple sensors. All the data transmission and connection between CNC machine, host server and AR devices are based on common TCP/IP (Transmission Control Protocol/Internet Protocol). Apart from collecting data from the physical machine tools, in order to get the virtual representation of its DT, an industrial camera has been installed inside the machine tool. The camera is used for real-time video capturing, then a Raspberry Pi microcomputer was used to receive the video stream and transmit it wirelessly from within the machine tool to remote display devices, in this case the touch-screen computer. For integrating the functional modules into the Intelligent Window as well as providing intuitive human-machine interactions, a composite Graphical User Interface (GUI), written in C# with Visual Studio, was developed as the user interface for the Intelligent Window.

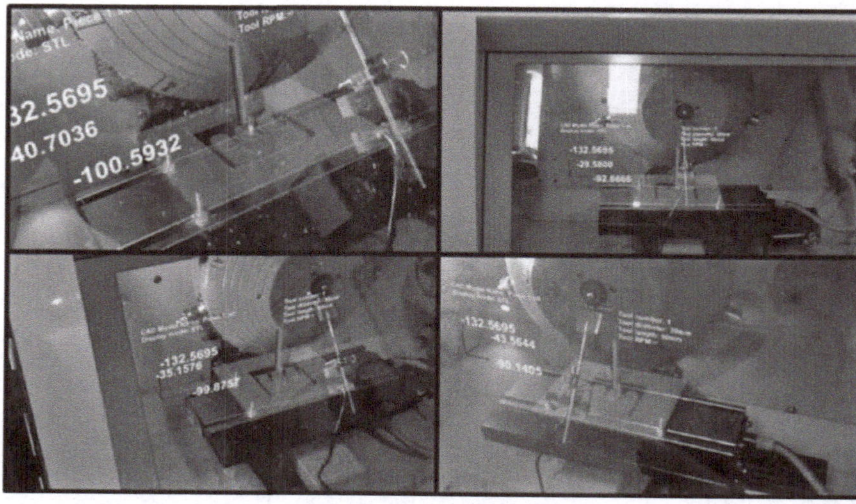

Figure 3.15 AR-enabled process monitoring and simulation.

This module integrates AR technology with collected DT data to provide an intuitive perception of machining process to operators. The DT data are processed and rendered on the video stream, including virtual cutter, workpieces, and real-time tool path, as shown in Figure 3.15. These virtual representations of the physical machine tools can help operators to clearly see the machining process even though the coolant and flying chips are blocking the actual view. Various alarms, warnings, and safety working volumes coming from the DT data or defined by the operators are rendered and shown through AR technology. Then the operator can quickly and easily distinguish errors message, and the physical machine tool can stop on time, to avoid any unexpected damage and mistakes. Besides, prognostic and health information can also be rendered on the related components to indicate the health status and remaining life so that proactive maintenance can be realized during the machining process.

This module also enables high-fidelity machining simulation based on the real machining environment with its DT data. An operator can do the simulation based on the physical machine tool under this mode. During the simulation process, real-time CNC feedback-based tracking ensures the accuracy and update speed of the positions of these virtual objects. Lastly, material removal process and collision detection are performed by calculating the intersections between the virtual cutter and other virtual objects. When the virtual cutter touches the unsafe area defined by the user, a collision alarm will appear, and the Emergency Stop will be automatically triggered.

3.6. Digital Twin-based robots

Small and medium-sized enterprises (SMEs) across the world still lack the confidence and knowledge to implement elements of Industry 4.0, such as DT technologies. One of the listed reasons is the desire to avoid high-investment costs. A possible solution is to use an effective retrofitting approach. At the beginning of every DT is the communication between the physical entity and its digital counterpart. Its most fundamental manifestation being the data collection. Further steps like data analysis, system control, and system optimization are taken from there.

3.6.1 Cost-effective connectivity tools

An enabler for cost-effective data collection solutions for KUKA robots is the software KukaVarProxy (KVP) (ImtsSrl/KUKAVARPROXY, 2017/2019). It was developed by the Italian company IMTS (Marking systems and Tagging machines for metallurgical products, n.d.). It has been available as freeware for a decade and was recently released under an open-source license (We are open-sourcing KUKAVARPROXY, 2018). This third-party software exposes a TCP/IP interface to read and write variables in the robot control system. While the software package sold by the robot manufacturer, called KUKA Robot Sensor Interface (RSI), can provide higher performance in advanced scenarios, e.g., hard real-time, the key advantage of KVP is that it is free and simple to use. It is compatible with all generations of the KUKA control systems, making it the first choice for retrofitting KUKA robots. Many research projects are based on KVP.

The first example of KVP usage is JOpenShowVar, developed by researchers from Norway together with IMTS (Sanfilippo, Hatledal, Zhang, Fago, & Pettersen, 2014, 2015). It acts as a middleware between KVP and user programs written in Java. As such, developers do not need to implement methods to create, send, and receive KVP message structures, but can use the methods provided by the library. On top of JOpenShowVar said researchers developed multiple case studies: an Android program to control the robot (Figure 3.16), line following with machine vision, control via Leap Motion controller, and controlling the robot through a haptic device (force-feedback joystick). The client program using JOpenShowVar must be on the same network as KVP.

3.6.2 Software suites for robotics

Another example of the KVP usage is a connectivity plug-in for the Robot Operating System (ROS) (Quigley et al., n.d.), developed by Mathias Arbo (*Itk-thrivaldi/kuka_kvp_hw_interface*, 2017/2019). It is an alternative to the also existing KUKA RSI connectivity plug-in. ROS is the main

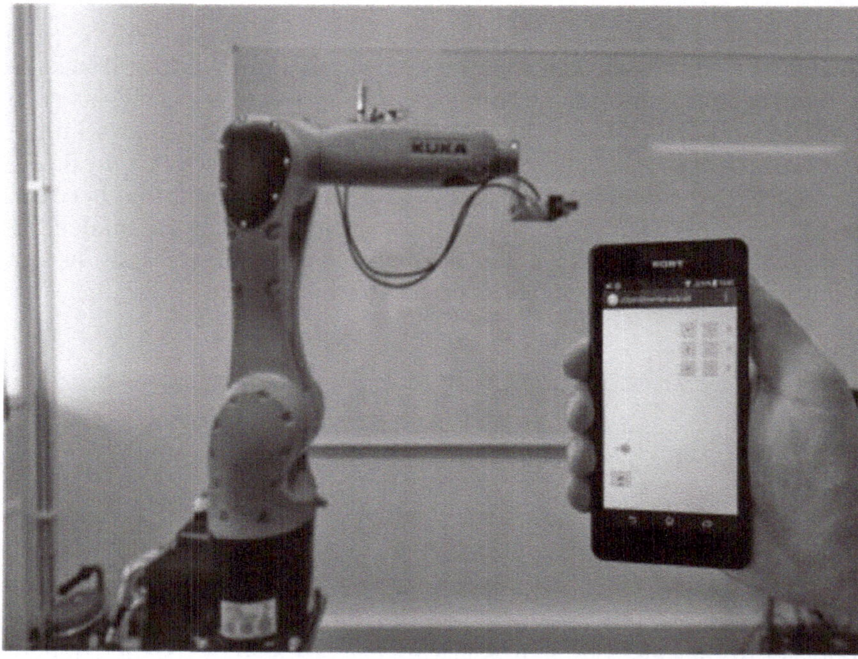

Figure 3.16 JOpenShowCar Case study: Kuka KR 6 R900 SIXX manipulator and GUI of Android mobile application used to control the arm. (Adapted from Sanfilippo et al., 2015.)

software tool for researchers working in robotics. It allows to model, simulates, control, and supervise complex robotic systems. It is used during development and actual field usage. As such ROS can be seen as the main platform for advanced DTs in robotics. Even though ROS was first released in 2007, making it older than the term DT. ROS and its plug-ins are open-source and mostly developed in C++, but user programs and plug-ins can also be written in Python.

Another example of a robot modeling and control software suite is RoboDK (Simulator for industrial robots and offline programming—RoboDK, n.d.), which is aimed at industrial users. RoboDK can connect and control a KUKA robot via KVP as well.

3.6.3 Case study: Cloud-based remote control

In similar fashion to the case studies from Norway, the LISMS research group at the University of Auckland retrofitted a KUKA KR-16/KR C2 industrial robot with a wireless connection to a cloud service. The project was implemented in two stages: in the first stage the data was only collected from the robot; in the second stage an additional control signal was sent to the robot enabling a remote user to supervise and control the robot. In contrast to the JOpenShowVar case studies, the client devices of this case study can communicate with the robot via the internet because a central messaging server, accessible from the internet, is used.

Figure 3.17 displays the structure of the case study: KukaVarProxy was deployed on the Windows-based KR C2 control system. Software developed by the research group and deployed to the control system, connected to this interface, read out the position data of the robot, and forwarded it via the control system's Serial Port to an attached microcontroller. The specially developed firmware connected via WiFi and the Internet to an MQTT messaging server running in the cloud and transmitted the position data. This constitutes the data collection: the data was brought into the cloud-service and could be further processed from here on. In this

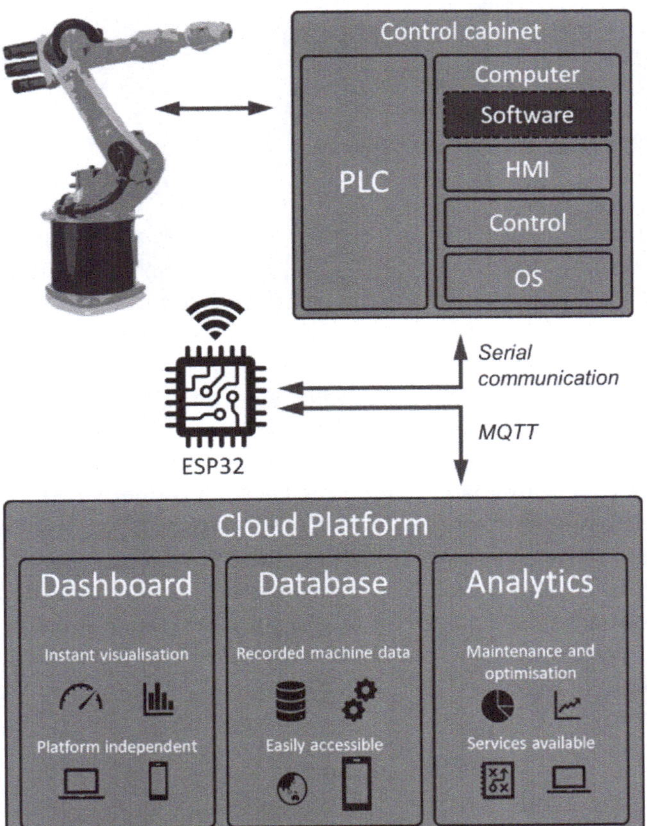

Figure 3.17 Structure of the case study and possible applications.

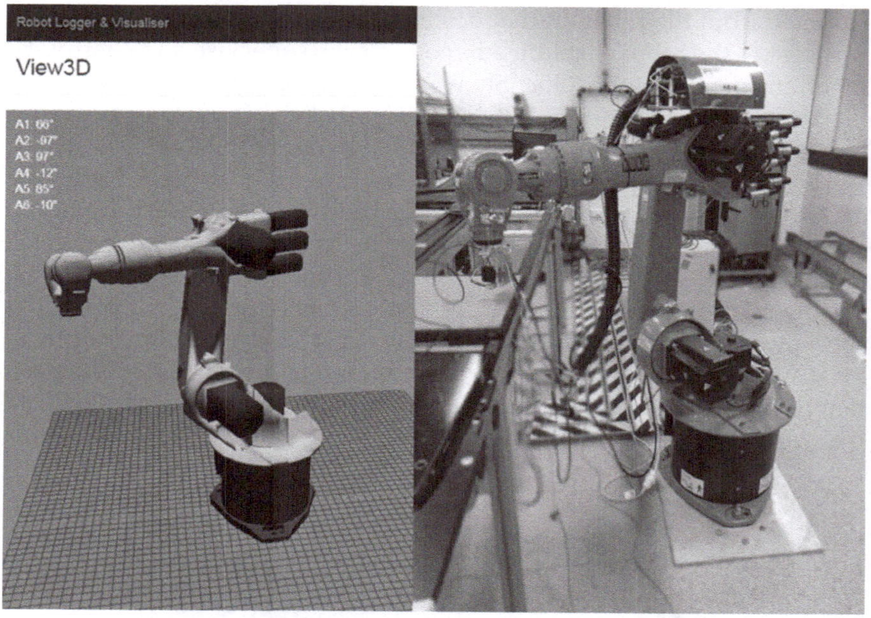

Figure 3.18 Real-time 3D model of the robot as displayed on a mobile client.

particular instance it was passed on to a web server serving a real-time 3D visualization and usage statistics to clients, mobile devices included (Figure 3.18). The reason to use the Serial Port combined with the microcontroller instead of using the control system's Ethernet port directly is that the port was already in use, which might also be the case in existing workshops, and the microcontroller's wireless modem eliminates the need for another network cable and socket.

In the second stage a client connected to the web server sent data from its joystick/gamepad through the HTML5 Gamepad API. The user's commands were forwarded to the custom software running on the control system, on its way passing through the messaging server, Internet, microcontroller, and Serial Port. The custom software sent data to KVP thus updating the target position in the control system, causing the robot to move. Combined with the 3D visualization and a streaming webcam, users were able to see the robot and move the robot effector through an obstacle course.

3.6.4 Conclusion

Although the presented research focused on KUKA industrial robots, it is still applicable to other industrial robots. ROS offers connectivity plug-ins for a wide range of devices and components. The solutions for the remote supervision and control of a robot can be used with any robot that supports some form of data collection or communication. Retrofitting is a cost-effective option for companies wishing to start implementing Industry 4.0 principles in their factories. We believe that the showcases and research are showing a trend towards easily deployable, yet complex, DTs of robots and other manufacturing equipment. This is achieved by modular middleware and connectivity plug-ins that allow manufacturer-agnostic integration into complex and powerful modeling and control software such as ROS or RoboDK. Utilizing internet connectivity of future or retrofitted devices the developers of control software could explore the possibility of changing their sales model from a computer-based license to a subscription service.

3.7 *Digital Twin-based process automation*

There is worldwide pressure to the process industry to continuously reduce production costs, increase product quality, and improve productivity, all at the same time (see, e.g., Poelvoorde, Eggert, & Vermeij, n.d.). The only way to handle this challenging situation is to optimize the production processes and their logistics in the best possible way, taking into account all the constraints coming from the "brownfield" application. Regarding logistics, one important aspect is suitable production planning to guarantee the optimal material flow inside the company to use all resources in the best possible way. Therefore, the solution of complex optimization tasks is necessary, which is usually done in a centralized way a long time before the production itself (see, e.g., Shah & Ierapetritou, 2012) and fully automated production planning not yet common praxis (Uhlemann, Lehmann, & Steinhilper, 2017). The calculations therefore use simplified models to reduce computation time (Technology-based assistance system for production planning in stainless mills (TECPLAN)—Publications Office of the EU, n.d.). As an example, steel production is quite a complex process, and thus the above-mentioned, simplified models cannot include all constraints of real production because otherwise it would lead to a combinatorial explosion, and thus the production planning would deliver black box solutions that are difficult to verify.

Furthermore, the production scheduling calculated by this centralized approach only works properly if no disturbances occur. In case of process disturbances or strong deviations of product properties, the initial optimal solution is not valid anymore and manual troubleshooting combined with the search for a possible solution for the production-scheduling task has to be started.

3.7.1 Process automation with Digital Twins

To overcome the above-described limitations of existing systems, a concept based on DTs has been developed that allows

1. real-time and decentralized optimization for scheduling tasks including detailed process models and taking into account all relevant properties of produced materials,
2. automatic self-optimization of the involved processes by using all relevant information coming from the process itself and coming from the produced material,
3. autonomous communication between plants, e.g., to inform the following process about plant disturbances or about deviating properties of incoming material.

The presented concept uses the idea of decentral interconnected process optimization and intensive communication with "intelligent" products, represented by its DTs. The concept is adapted to the special requirements and conditions of the steel industry for the process route from the hot strip mill over coil cooling to the pickling line.

Based on the main idea of enhancing the product intelligence, via DTs, the coils are uniquely identifiable and locatable at all times, if we know their historical process data and their detailed current status and properties, e.g., temperature distribution and scale thickness. The detailed current status and properties are simulated using suitable models. All coils allocate themselves for the next processes and transmit their actual properties and processing targets. As the coils cannot communicate themselves with their virtual representative, the DTs take over these jobs.

With the introduction of DTs for products and processes, as shown in Figure 3.19, autonomous communication becomes easier, e.g., for machine-to-machine communication (M2M) (Mehmood, Görg, Muehleisen, & Timm-Giel, 2015). In comparison to classical communication between machines, the challenges here exist, e.g., in the continually changing communication partners. For each new product, another DT is created, which is then a new communication partner. The DTs of the products have only a limited active life phase, i.e., only as long as they go through the production processes. They are archived afterward and removed from the active system. In addition, if the communication is event-based and not cyclical, the decision when a communication has to take place is made by the DT itself. Examples of these events are process disturbances, interruptions, detected systematic faults, and the finalization of a processing step.

Information from intelligent products/DTs and from M2M communication are used for process self-optimization. In this context, process self-optimization is about fine-tuning the setup parameters of the process, e.g.,

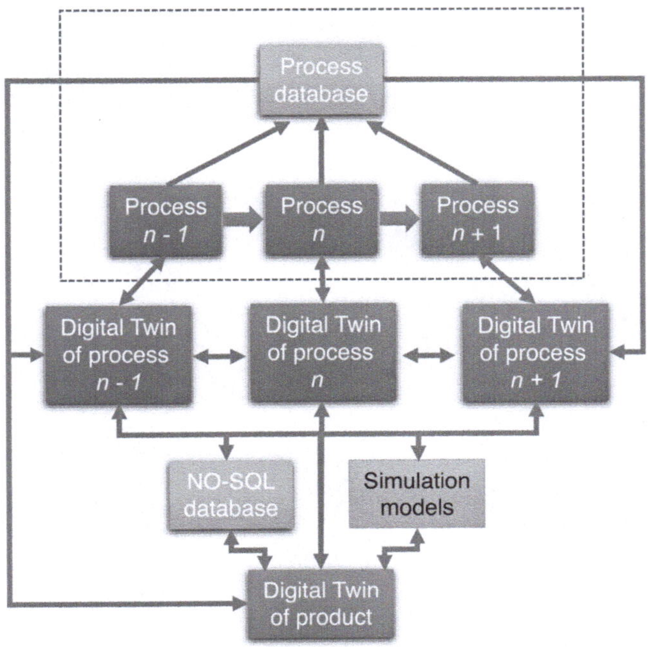

Figure 3.19 DTs for process optimization.

the plant speed, autonomously. The communicated individual properties of the products play a decisive role here. A further aspect is the optimization of the order in which the products are processed. Depending on the production process, a slight change in the production sequence of the products can lead to a significant increase in throughput, e.g., if shorter plant downtimes are the result. DTs of the process can determine the optimum production sequence of all available products using dynamic simulation models. A further aspect is systematic detected processing faults and process interruptions of the subsequent process that are transmitted to the previous one. Thus, the previous process utilizes this information to optimize its processing parameters.

The following industrial case study explains in detail how the autonomous communication works and shows the concept of DTs for process optimization.

3.7.2 *Case study: Throughput optimization in a steel mill with Digital Twins*

3.7.2.1 *Process description*

The overview in Figure 3.20 shows parts of the steel production (from reheating furnace of the hot strip mill to the pickling line) and the communication

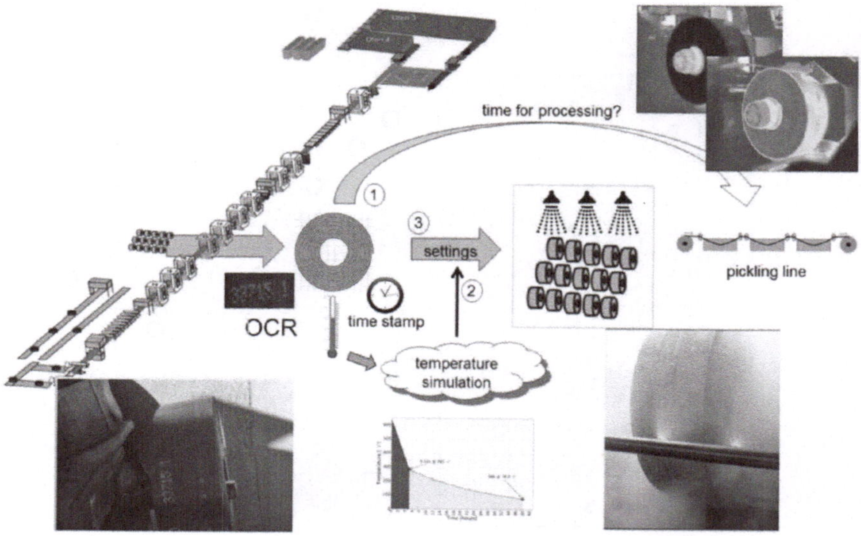

Figure 3.20 DTs for throughput optimization in a steel mill.

steps of the DTs. When producing a steel strip, the slabs (thickness approx. 26 cm) are heated in the reheating furnace for better formability. After the slab has been heated to approx. 1250 degrees (red glowing), several rolling stands reduce the thickness (typically to approx. 2–6 mm) and slightly adjust the strip width. The subsequent cooling section (run out table) significantly determines the material properties of the metal strip. The coiler (at the end of the cooling section) rolls the strip into a so-called coil, which at this time has a temperature of approx. 500–800 degrees.

When steel is scorching, scale quickly builds upon the surface. To prevent scale growth as fast as possible, there is a coil shower in this plant that rapidly lowers the coil temperature to room temperature. Forklift trucks transport the coils from the hot strip mill to the coil shower. The time required for this varies and depends on the utilization of the forklift trucks.

The subsequent process stage, the pickling line, removes the scale that always forms due to the process. In the pickling line, the steel strip passes through several tanks filled with acid. After pickling, the strip is rolled up again into a coil and leaves the factory carried by a truck.

Given task: Increasing the pickling line throughput

3.7.2.2 *The solution from the process point of view*

If the coil would enter the pickling line warm instead of at outside temperature, the acid of the pickling line has a stronger effect and shorter residence time in the acid tanks is sufficient. This means that the strip speed can be higher in the pickling line. A further aspect of increasing strip speed is the knowledge about the thickness of the scale on the strip. With a low scale buildup, the strip can also be run through the pickling line more quickly because less scale has to be removed. This information is usually not available to the plant operator and changes frequently during production. The decisive factor here is the transport time from the recoiler of host strip mill to the coil shower. The longer this process takes, the more scale grows during this time, as the coil is still very hot.

Constraints

- The desired strip temperature at the end of the cooling section of the hot strip mill depends on the product and usually lies between 500 and 800 degrees.
- The transport time from the hot rolling mill to the coil shower varies.
- The weather (outside temperature and rain) is variable and only predictable relatively inaccurately.
- The exact processing time in the subsequent process (pickling) is not known at an early stage and can also change; e.g., in the event of an unplanned plant downtime due to a strip break.

3.7.2.3 Solution

Usually, the machine knows best how to process the product. Since there are many uncertain parameters, such as transport time, weather, or the time of further processing, a decentralized decision strategy for the determination of the process parameters is a very good alternative in this case.

The product, in this case, the coil, cannot communicate with the plant. You, therefore, need an Adaptive Digital Twin that can perform this task for the coil.

The Adaptive Digital Twin can communicate with other DTs and has access to stored process data of previous production steps. The Adaptive Digital Twin also knows the quality requirements of the assigned customer.

DTs can represent products, i.e. the coils as well as processes, e.g., the pickling line. Furthermore, DTs can run simulations with models to calculate nonmeasurable properties such as scale growth or strip temperature. Since DTs can also communicate with any other process, it is possible to query the current weather and a short-term forecast.

The DTs are thus able to autonomously determine the optimal process parameters and make decisions within a defined framework.

Workflow

The birth of the DT of the coil happens, when the coil is transported from the hot strip mill to the coil shower. Therefore a camera has been installed which detects the coil number with OCR-software. The time of completion the coil in the hot strip mill is known and stored in the process database. This allows the total time between completion in the hot strip mill and the start of water cooling in the coil shower to be determined. A new Adaptive Digital Twin is a new virtual object, with predefined algorithms for behavior and communication.

The first task of our newly born Adaptive Digital Twin of the coil is to ask the Adaptive Digital Twin of the pickling line for its planned processing time in the pickling line:

- Adaptive Digital Twin of Coil to Adaptive Digital Twin of pickling line: "I am Coil 4711, steel grade x, thickness y, width z. When is my planned processing time in the pickling line and what temperature should I have?"
- The Adaptive Digital Twin of the pickling line calculates the production time of all coils already registered in order to respond with the best possible forecast.
- Adaptive Digital Twin of the pickling line to Adaptive Digital Twin of Coil: "Planned processing time: 18 h, max; incoming temperature: 80 degrees."

The Adaptive Digital Twin of the coil still lacks further information before the optimum cooling time in the coil shower can be calculated, which is why the next two communication steps are the following:

- The Adaptive Digital Twin of the coil queries in the process database its strip temperature at the end of the hot strip mill and the current weather situation at the weather station.

Now, all necessary information for the calculation of the optimal water cooling time is available:

- The Adaptive Digital Twin of the coil starts the model calculation for the water cooling process in the coil shower with the parameters:
- Further processing in the pickling line in, e.g., 18 h
- Current temperature, e.g., 650 degrees, target temperature in 18 h, e.g., 80 degrees
- Outdoor temperature, e.g., 10 degrees, no rain, temperature of cooling water
- Coil geometry (thickness, width, length)

The calculation result is the optimal duration time of active water cooling, e.g., only 4 h, since 14 h follow at 10 degrees outdoor temperature.

The Adaptive Digital Twin now knows the optimum process parameter for the next process step, the coil shower, and must communicate this.

- Adaptive Digital Twin of the coil to coil shower: "I am Coil 4711, please only 4 h active water cooling."

The Adaptive Digital Twin of coil now calculates the scale growth based on the temperature evolution and the observed transport time from the hot strip mill to the coil shower.

After the active water cooling the coil is transported from the coil shower to the pickling line. Once there, the Adaptive Digital Twin now has important information about its condition that was previously unknown:

- Temperature curve over the strip length and a recalculation into strip speed offsets.
- Simulated scale thickness on the strip also recalculation into strip speed offsets.

When the coil is inserted into the pickling line, there is again communication between the Adaptive Digital Twin of the coil and the Adaptive Digital Twin of the pickling line.

- Adaptive Digital Twin of the coil to Adaptive Digital Twin of pickling line: "I am Coil 4711 and have the attached temperature curve and speed offset curve. Add the speed offset to the standard speed. My scale thickness is only y μm, i.e., I can additionally be produced z m/min faster than the standard speed."

Again the product, i.e., the coil, tells the plant, i.e., the pickling line, about the fine-tuning of the speed setup.

3.7.2.4 Industrial results

The industrial testing of the DTs concept has produced several and partly surprising results:

- As expected, a significant increase in the pickling throughput could be demonstrated. For the two steel grades initially investigated, the increase in speed was 16% and 20%, respectively, which is a considerable increase in performance, especially considering that no changes were made to the pickling process or the pickling tanks themselves.
- The coils produced using the new method arrive warm in the pickling line. This considerably reduces the energy required to heat the pickling tanks, since the coils no longer have to be heated from the outside temperature (often 0 degrees in winter) to the pickling temperature of over 80 degrees. Each coil has a weight of approx. 10,000 kg.
- An unexpectedly positive effect of the warm coils concerns the elasticity of the steel. For a particularly critical steel grade, e.g., it was observed that the risk of strip breaks in the event of damage to the strip edge was considerably reduced. This significantly increases the production time of the pickling line, since the repair

of a strip break often takes several hours and the line cannot produce during this time. Any strip breakage that is prevented therefore immediately means several hours of "additional" production time and at no extra cost.

The use of DTs in this steel plant leads to a significant increase in productivity, energy savings, and less unscheduled plant downtime.

3.8 Discussions and conclusions

This chapter discussed the vision of DT-enabled smart manufacturing in the era of Industry 4.0. Key concepts of smart manufacturing and DT were discussed substantially; in particular the changes and opportunities that DT can bring to design and manufacturing operations and systems are discussed in detail with case studies from the Laboratory for Industry 4.0 Smart Manufacturing Systems, University of Auckland.

Smart manufacturing is believed to be the next generation of manufacturing systems that features autonomous machine-to-machine communication, reconfigurable manufacturing processes, self-optimization decision-making, and end-to-end traceability. DT, as an emerging technology for replicating physical assets in the cyberspace offers tremendous potentials for asset monitoring, cloud-based decision-making, and AI-augmented decision-making. The presented case studies showcased principles on using DT technologies to make product design process smarter, in particular for personalized smart products. We have also developed DT-based machine tools as smart machine tools that offer new value-added services, such as AR-based visualization and monitoring, and predictive maintenance. Similarly, DT-based industrial robots provide operations options to remote monitor and maximize uptime of expensive assets. When DTs are deployed to a production line, we can utilize the ubiquitous connectivity and smart decision-making capabilities of key system components to achieve decentralized decision-making for throughput optimization.

We believe the rapid development of DT technologies can significantly change the operations and value creations of design and manufacturing in the future. Future research should focus on the advancement of core DT technologies, including the following.

First, there is a lack of holistic approaches to developing a DT that incorporates top-down standardization needs and bottom-up application motivations. DT is application-driven; the model of a DT is determined by the application scenarios of the DT; therefore, the model and architecture can vary. However, standardization is of great importance to achieve ubiquitous connectivity and communications between a DT and its physical counterpart, which requires a flexible system framework and architecture that is compatible with existing networking and communication standards. Research on integrating top-down standardization approach and bottom-up application-driven DT development is urgently needed.

Second, data science for developing DTs is missing. We need breakthrough in determining what type of data to be collected at what time, with what frequency, and for what purpose for developing a successful DT. Third, there is no systematic way to verify and evaluate the accuracy and performance of a DT. DTs can be developed without optimization in mind. The research currently lacks a critical evaluation framework to testify the quality of a developed DT, which can potentially cause significant DT development and operation costs.

Fourth, there is no established generic way of monetizing DT services. DTs create massive opportunity for business transformation; DTs are used either for internal business improvement or for high-value product offerings. Monetizing DT and its associated services is critical to business growth in New Zealand. Unfortunately, limited research has been undertaken in identifying business opportunities and developing the enabling commercialization pathway for DT-related products and services.

References

Angrish, A., Starly, B., Lee, Y.-S., & Cohen, P. H. (2017). A flexible data schema and system architecture for the virtualization of manufacturing machines (VMM). Journal of Manufacturing Systems, 45, 236–247. https://doi.org/10.1016/j.jmsy.2017.10.003

Augment. (2016, May 12). Infographic: The History of Augmented Reality. Retrieved 20 October 2019, from Augment News website: https://www.augment.com/blog/infographic-lengthy-history-augmented-reality/

Berg, L. P., & Vance, J. M. (2017). Industry use of virtual reality in product design and manufacturing: A survey. Virtual Reality, 21(1), 1–17. https://doi.org/10.1007/s10055-016-0293-9

Bielefeldt, B., Hochhalter, J., & Hartl, D. (2015, September 21). Computationally Efficient Analysis of SMA Sensory Particles Embedded in Complex Aerostructures Using a Substructure Approach. V001T02A007-V001T02A007: https://doi.org/10.1115/SMASIS2015-8975

Bolton, R. N., McColl-Kennedy, J. R., Cheung, L., Gallan, A., Orsingher, C., Witell, L., & Zaki, M. (2018). Customer experience challenges: Bringing together digital, physical and social realms. Journal of Service Management. https://doi.org/10.1108/JOSM-04-2018-0113

Booth, J. W., Alperovich, J., Chawla, P., Ma, J., Reid, T. N., & Ramani, K. (2017). The Design for additive manufacturing worksheet. Journal of Mechanical Design, 139(10). https://doi.org/10.1115/1.4037251

Bordegoni, M., Cugini, U., Caruso, G., & Polistina, S. (2009). Mixed prototyping for product assessment: A reference framework. International Journal on Interactive Design and Manufacturing (IJIDeM), 3(3), 177–187. https://doi.org/10.1007/s12008-009-0073-9

Boschert, S., & Rosen, R. (2016). Digital twin—the simulation aspect. In P. Hehenberger & D. Bradley (Eds.), Mechatronic Futures: Challenges and Solutions for Mechatronic Systems and their Designers (pp. 59–74). https://doi.org/10.1007/978-3-319-32156-1_5

Buchholz, C., Vorsatz, T., Kind, S., & Stark, R. (2017). SHPbench – a smart hybrid prototyping based environment for early testing, verification and (user based) validation of advanced driver assistant systems of cars. Procedia CIRP, 60, 139–144. https://doi.org/10.1016/j.procir.2017.02.025

Bukkapatnam, S., & Clark, B. (2007). Dynamic modeling and monitoring of contour crafting—an extrusion-based layered manufacturing process. Journal of Manufacturing Science and Engineering, 129(1), 135–142. https://doi.org/10.1115/1.2375137

Camburn, B., Viswanathan, V., Linsey, J., Anderson, D., Jensen, D., Crawford, R., & Wood, K. (2017). Design prototyping methods: State of the art in strategies, techniques, and guidelines. Design Science, 3. https://doi.org/10.1017/dsj.2017.10

Catarci, T., Firmani, D., Leotta, F., Mandreoli, F., Mecella, M., & Sapio, F. (2019). A conceptual architecture and model for smart manufacturing relying on service-based digital twins. 2019 IEEE International Conference on Web Services (ICWS), 229–236. https://doi.org/10.1109/ICWS.2019.00047

Ceruti, A., Liverani, A., & Bombardi, T. (2017). Augmented vision and interactive monitoring in 3D printing process. International Journal on Interactive Design and Manufacturing (IJIDeM), 11(2), 385–395. https://doi.org/10.1007/s12008-016-0347-y

Cheng, T., Zhang, J., Hu, C., Wu, B., & Yang, S. (2001). Intelligent machine tools in a distributed network manufacturing mode environment. The International Journal of Advanced Manufacturing Technology, 17(3), 221–232. https://doi.org/10.1007/s001700170194

Cheng, Y., & Jafari, M. A. (2008). Vision-based online process control in manufacturing applications. IEEE Transactions on Automation Science and Engineering, 5(1), 140–153. https://doi.org/10.1109/TASE.2007.912058

Delli, U., & Chang, S. (2018). Automated process monitoring in 3D printing using supervised machine learning. Procedia Manufacturing, 26, 865–870. https://doi.org/10.1016/j.promfg.2018.07.111

Digital Twin | Siemens. (n.d.). Retrieved 1 September 2019, from Siemens PLM Software website: https://www.plm.automation.siemens.com/global/en/our-story/glossary/digital-twin/24465

Eiriksson, E. R., Pedersen, D. B., Frisvad, J. R., Skovmand, L., Heun, V., Maes, P., & Aanæs, H. (2017). Augmented reality interfaces for additive manufacturing. In P. Sharma & F. M. Bianchi (Eds.), Image Analysis (pp. 515–525). New York, NY: Springer International Publishing.

Evans, P. L., Castellazzi, A., & Johnson, C. M. (2016). Design tools for rapid multidomain virtual prototyping of power electronic systems. IEEE Transactions on Power Electronics, 31(3), 2443–2455. https://doi.org/10.1109/TPEL.2015.2437793

Fang, T., Jafari, M. A., Bakhadyrov, I., Safari, A., Danforth, S., & Langrana, N. (1998). Online defect detection in layered manufacturing using process signature. SMC'98 Conference Proceedings. 1998 IEEE International Conference on Systems, Man, and Cybernetics (Cat. No.98CH36218), 5, 4373–4378. https://doi.org/10.1109/ICSMC.1998.727536

Holzmond, O., & Li, X. (2017). In situ real time defect detection of 3D printed parts. Additive Manufacturing, 17, 135–142. https://doi.org/10.1016/j.addma.2017.08.003

Hoover, M. (2018). An evaluation of the Microsoft HoloLens for a manufacturing-guided assembly task. Graduate Theses and Dissertations. https://doi.org/10.31274/etd-180810-6008

ImtsSrl/KUKAVARPROXY [Visual Basic]. (2019). Retrieved from: https://github.com/ImtsSrl/KUKAVARPROXY (Original work published 2017)

Itk-thrivaldi/kuka_kvp_hw_interface [C++]. (2019). Retrieved from: https://github.com/itk-thrivaldi/kuka_kvp_hw_interface (Original work published 2017)

Jin, Z., Zhang, Z., & Gu, G. X. (2019). Autonomous in-situ correction of fused deposition modeling printers using computer vision and deep learning. Manufacturing Letters, 22, 11–15. https://doi.org/10.1016/j.mfglet.2019.09.005

julieseto. (n.d.). Overview of Azure Digital Twins. Retrieved 1 September 2019, from: https://docs.microsoft.com/en-us/azure/digital-twins/about-digital-twins

Kagermann, H., Helbig, J., Hellinger, A., & Wahlster, W. (2013). Recommendations for Implementing the Strategic Initiative INDUSTRIE 4.0: Securing the Future of German Manufacturing Industry; Final Report of the Industrie 4.0 Working Group. Forschungsunion.

Kim, B., Kashiba, Y., Dai, S., & Shiraishi, S. (2017). Testing autonomous vehicle software in the virtual prototyping environment. IEEE Embedded Systems Letters, 9(1), 5–8. https://doi.org/10.1109/LES.2016.2644619

Kim, S., Rosen, D. W., Witherell, P., & Ko, H. (2019). A design for additive manufacturing ontology to support manufacturability analysis. Journal of Computing and Information Science in Engineering, 19(4). https://doi.org/10.1115/1.4043531

Kusiak, A. (2018). Smart manufacturing. International Journal of Production Research, 56(1–2), 508–517. https://doi.org/10.1080/00207543.2017.1351644

Lanzotti, A., Carbone, F., Grazioso, S., Renno, F., & Staiano, M. (2018). A new interactive design approach for concept selection based on expert opinion. International Journal on Interactive Design and Manufacturing (IJIDeM), 12(4), 1189–1199. https://doi.org/10.1007/s12008-018-0482-8

Lasi, H., Fettke, P., Kemper, H.-G., Feld, T., & Hoffmann, M. (2014). Industry 4.0. Business & Information Systems Engineering, 6(4), 239–242. https://doi.org/10.1007/s12599-014-0334-4

Liu, C., Vengayil, H., Zhong, R. Y., & Xu, X. (2018). A systematic development method for cyber-physical machine tools. Journal of Manufacturing Systems, 48, 13–24. https://doi.org/10.1016/j.jmsy.2018.02.001

Liu, C., & Xu, X. (2017). Cyber-physical machine tool – the era of machine tool 4.0. Procedia CIRP, 63, 70–75. https://doi.org/10.1016/j.procir.2017.03.078

Liu, C., Xu, X., Peng, Q., & Zhou, Z. (2018). MTConnect-based cyber-physical machine tool: A case study. Procedia CIRP, 72, 492–497. https://doi.org/10.1016/j.procir.2018.03.059

Lu, Y., Liu, C., Wang, K. I.-K., Huang, H., & Xu, X. (2020). Digital twin-driven smart manufacturing: Connotation, reference model, applications and research issues. Robotics and Computer-Integrated Manufacturing, 61, 101837. https://doi.org/10.1016/j.rcim.2019.101837

Lu, Y., & Xu, X. (2018a). A Digital Twin Reference Model for Smart Manufacturing. 2018-December. Retrieved from Scopus.

Lu, Y., & Xu, X. (2018b). Resource virtualization: A core technology for developing cyber-physical production systems. Journal of Manufacturing Systems, 47, 128–140. https://doi.org/10.1016/j.jmsy.2018.05.003

Luo, W., Hu, T., Zhang, C., & Wei, Y. (2019). Digital twin for CNC machine tool: Modeling and using strategy. Journal of Ambient Intelligence and Humanized Computing, 10(3), 1129–1140. https://doi.org/10.1007/s12652-018-0946-5

Madni, A. M., Madni, C. C., & Lucero, S. D. (2019). Leveraging digital twin technology in model-based systems engineering. Systems, 7(1), 7. https://doi.org/10.3390/systems7010007

Makris, S., Karagiannis, P., Koukas, S., & Matthaiakis, A.-S. (2016). Augmented reality system for operator support in human–robot collaborative assembly. CIRP Annals, 65(1), 61–64. https://doi.org/10.1016/j.cirp.2016.04.038

Marking systems and Tagging machines for metallurgical products. (n.d.). Retrieved 20 October 2019, from Imts website: https://www.imts.eu/

Mars, W. V., Suter, J. D., & Bauman, M. (2018, April 3). Computing Remaining Fatigue Life Under Incrementally Updated Loading Histories. 2018-01–0623: https://doi.org/10.4271/2018-01-0623

Mehmood, Y., Görg, C., Muehleisen, M., & Timm-Giel, A. (2015). Mobile M2M communication architectures, upcoming challenges, applications, and future directions. EURASIP Journal on Wireless Communications and Networking, 2015(1), 250. https://doi.org/10.1186/s13638-015-0479-y

MuchMoreThanDNC.pdf. (n.d.). Retrieved from: http://www.machinery.co.uk/article-images/3532/MuchMoreThanDNC.pdf

Negri, E., Fumagalli, L., & Macchi, M. (2017). A review of the roles of digital twin in CPS-based production systems. Procedia Manufacturing, 11, 939–948. https://doi.org/10.1016/j.promfg.2017.07.198

Neumann, P. (2007). Communication in industrial automation—what is going on? Control Engineering Practice, 15(11), 1332–1347. https://doi.org/10.1016/j.conengprac.2006.10.004

Parise, S., Guinan, P. J., & Kafka, R. (2016). Solving the crisis of immediacy: How digital technology can transform the customer experience. Business Horizons, 59(4), 411–420. https://doi.org/10.1016/j.bushor.2016.03.004

Pellicciari, M., Berselli, G., & Balugani, F. (2015). On designing optimal trajectories for servo-actuated mechanisms: Detailed virtual prototyping and experimental evaluation. IEEE/ASME Transactions on Mechatronics, 20(5), 2039–2052. https://doi.org/10.1109/TMECH.2014.2361759

Poelvoorde, G., Eggert, A., & Vermeij, J. (n.d.). Annual Report 2019. Retrieved 20 October 2019, from EUROFER, The European Steel Association website: http://www.eurofer.org/News%26Events/PublicationsLinksList/201907-AnnualReport.pdf

Qi, Q., & Tao, F. (2018). Digital twin and big data towards smart manufacturing and industry 4.0: 360 degree comparison. IEEE Access, 6, 3585–3593. https://doi.org/10.1109/ACCESS.2018.2793265

Quigley, M., Gerkey, B., Conley, K., Faust, J., Foote, T., Leibs, J., & Ng, A. (n.d.). ROS: An Open-Source Robot Operating System. 6.

Revetria, R., Tonelli, F., Damiani, L., Demartini, M., Bisio, F., & Peruzzo, N. (2019). A real-time mechanical structures monitoring system based on digital twin, iot and augmented reality. 2019 Spring Simulation Conference (SpringSim), 1–10. https://doi.org/10.23919/SpringSim.2019.8732917

Sanfilippo, F., Hatledal, L. I., Zhang, H., Fago, M., & Pettersen, K. Y. (2014). JOpenShowVar: An open-source cross-platform communication interface to Kuka robots. 2014 IEEE International Conference on Information and Automation (ICIA), 1154–1159. https://doi.org/10.1109/ICInfA.2014.6932823

Sanfilippo, F., Hatledal, L. I., Zhang, H., Fago, M., & Pettersen, K. Y. (2015). Controlling kuka industrial robots: Flexible communication interface JOpenShowVar. IEEE Robotics Automation Magazine, 22(4), 96–109. https://doi.org/10.1109/MRA.2015.2482839

Shafto, M., Conroy, M., Doyle, R., Glaessgen, E., Kemp, C., LeMoigne, J., & Wang, L. (2012). Modeling, simulation, Information Technology & Processing Roadmap. National Aeronautics and Space Administration.

Shah, N. K., & Ierapetritou, M. G. (2012). Integrated production planning and scheduling optimization of multisite, multiproduct process industry. Computers & Chemical Engineering, 37, 214–226. https://doi.org/10.1016/j.compchemeng.2011.08.007

Simulator for industrial robots and offline programming—RoboDK. (n.d.). Retrieved 20 October 2019, from: https://robodk.com/index

Tan, Y., Yang, W., Yoshida, K., & Takakuwa, S. (2019). Application of IoT-Aided simulation to manufacturing systems in cyber-physical system. Machines, 7(1), 2. https://doi.org/10.3390/machines7010002

Tao, F., Zhang, H., Liu, A., & Nee, A. Y. C. (2019). Digital twin in industry: State-of-the-art. IEEE Transactions on Industrial Informatics, 15(4), 2405–2415. https://doi.org/10.1109/TII.2018.2873186

Tao, Fei, Zhang, M., & Nee, A. Y. C. (2019). Chapter 1—background and concept of digital twin. In Fei Tao, M. Zhang, & A. Y. C. Nee (Eds.), Digital Twin Driven Smart Manufacturing (pp. 3–28). https://doi.org/10.1016/B978-0-12-817630-6.00001-1

Technology-based assistance system for production planning in stainless mills (TECPLAN)—Publications Office of the EU. (n.d.). Retrieved 20 October 2019, from: https://op.europa.eu/en/publication-detail/-/publication/f0f872f5-e432-11e5-8a50-01aa75ed71a1

Thoben, K.-D., Wiesner, S., & Wuest, T. (2017). "Industrie 4.0" and Smart Manufacturing – A Review of Research Issues and Application Examples. 14.

Thompson, K. D. (2014, April 25). Smart Manufacturing Operations Planning and Control Program. Retrieved 15 August 2019, from NIST website: https://www.nist.gov/programs-projects/smart-manufacturing-operations-planning-and-control-program

Tuegel, E. (2012). The airframe digital twin: Some challenges to realization. 53rd AIAA/ASME/ASCE/AHS/ASC Structures, Structural Dynamics and Materials Conference 20th AIAA/ASME/AHS Adaptive Structures Conference 14th AIAA, 1812.

Tuegel, E. J., Ingraffea, A. R., Eason, T. G., & Spottswood, S. M. (2011). Reengineering Aircraft Structural Life Prediction Using a Digital Twin [Research article]: https://doi.org/10.1155/2011/154798

Uhlemann, T. H.-J., Lehmann, C., & Steinhilper, R. (2017). The digital twin: Realizing the cyber-physical production system for industry 4.0. Procedia CIRP, 61, 335–340. https://doi.org/10.1016/j.procir.2016.11.152

Vergnano, A., Berselli, G., & Pellicciari, M. (2017). Parametric virtual concepts in the early design of mechanical systems: A case study application. International Journal on Interactive Design and Manufacturing (IJIDeM), 11(2), 331–340. https://doi.org/10.1007/s12008-015-0295-y

Wang, G. G. (2002). Definition and review of virtual prototyping. Journal of Computing and Information Science in Engineering, 2(3), 232–236. https://doi.org/10.1115/1.1526508

We are open-sourcing KUKAVARPROXY. (2018, January 8). Retrieved 20 October 2019, from Imts website: https://www.imts.eu/open-sourcing-kukavarproxy/

Weyer, S., Meyer, T., Ohmer, M., Gorecky, D., & Zühlke, D. (2016). Future modeling and simulation of CPS-based factories: An example from the automotive industry. IFAC-PapersOnLine, 49(31), 97–102. https://doi.org/10.1016/j.ifacol.2016.12.168

Xu, X. (2008). Integrating Advanced Computer-Aided Design, Manufacturing, and Numerical Control: Principles and Implementations. Hershey, PA: Information Science Reference - Imprint of: IGI Publishing.

Xu, X. (2017). Machine tool 4.0 for the new era of manufacturing. The International Journal of Advanced Manufacturing Technology, 92(5), 1893–1900. https://doi.org/10.1007/s00170-017-0300-7

Zheng, P., Yu, S., Wang, Y., Zhong, R. Y., & Xu, X. (2017). User-experience based product development for mass personalization: A case study. Procedia CIRP, 63, 2–7.

Zhong, R. Y., Xu, X., Klotz, E., & Newman, S. T. (2017). Intelligent manufacturing in the context of industry 4.0: A review. Engineering, 3(5), 616–630. https://doi.org/10.1016/J.ENG.2017.05.015

Zhou, K., Liu, T., & Zhou, L. (2015). Industry 4.0: Towards future industrial opportunities and challenges. 2015 12th International Conference on Fuzzy Systems and Knowledge Discovery (FSKD), 2147–2152. https://doi.org/10.1109/FSKD.2015.7382284

chapter four

Industry 4.0 in small- and medium-sized enterprises

A literature review on research clusters

Julian M. Müller
Salzburg University of Applied Sciences, Puch bei Hallein, Austria

Contents

4.1 Introduction

Industry 4.0, internationally also known as the Industrial Internet of Things, describes the digitally enabled horizontal and vertical interconnection of industrial value creation (Kagermann et al., 2013). The concept, describing an intended fourth industrial revolution, has gained large interest in managerial practice and academic research throughout several research disciplines. While technological research and showcases on Industry 4.0, often limited to single factories, are leading research and practical implementation, several topics are lagging behind and need to be addressed better, such as small- and medium-sized enterprises (SMEs). Especially relating to their importance in industrial supply chains, their integration within the concept of Industry 4.0 is of high importance (Kagermann et al., 2013; Kiel et al., 2017).

Despite their importance in industrial value creation, SMEs have been regarded considerably less in extant research on Industry 4.0

(Kiel et al., 2017). Comparably, the implementation of Industry 4.0 across the supply chain, often consisting of SMEs, remains little understood in comparison to other research areas of Industry 4.0 (Birkel et al., 2019; Veile et al., 2019a). In response to this research gap, and to facilitate future research on SMEs in the context of Industry 4.0, this chapter identifies, examines, and compares the most prominent research clusters of Industry 4.0 related research on SMEs.

4.2 Theoretical background

4.2.1 Industry 4.0

The term Industry 4.0 is derived from an expected industrial revolution, after three previous industrial revolutions. These include steam power in the first industrial revolution, electrification in the second industrial revolution, and automation in the third industrial revolution. While the previous industrial revolutions were described ex post, Industry 4.0 was proclaimed ex ante, following a concept of the German government. The term Industry 4.0 first occurred in 2011 at the Hanover Fair, describing a concept of the German government to safeguard the competitiveness of its manufacturing industry (Kagermann et al., 2013).

In its core, Industry 4.0 has its technological foundation in cyber-physical systems, allowing to merge the physical and virtual worlds. This requires sensors to replicate the physical world in the virtual one, data processing and analysis capabilities, and capabilities for communication and interaction (Kagermann et al., 2013; Kiel et al., 2017).

These technological foundations allow three central characteristics: horizontal interconnection, vertical interconnection, and end-to-end engineering. Horizontal interconnection refers to digital interconnection across company boundaries, i.e., the supply chain. Vertical interconnection describes the digital interconnection within enterprises, i.e., company functions. Further, end-to-end engineering describes interconnection from product development, production, usage, and recycling, i.e., the entire product life cycle. The digital interconnection of these dimension is enabled by cyber-physical systems, sensors, data processing and transmission capabilities, and thereupon data analysis techniques, as described above (Kagermann et al., 2013; Kiel et al., 2017). Notably, the different data standards, interfaces, and company boundaries represent a challenge for Industry 4.0 implementation. This is especially true for SMEs with limited resources, capabilities, and since SMEs fear dependencies and increased transparency. While the potentials offered through Industry 4.0 are predicted to bring several benefits to industrial value creation in economic, ecological, and social regards, several risks and barriers need to be overcome to implement the concept, especially across the supply chain, and in

SMEs (Birkel et al., 2019; Kiel et al., 2017; Moeuf et al., 2018; Müller, Buliga, & Voigt, 2018a; Veile et al., 2019a).

Worldwide, comparable concepts to Industry 4.0 have been initiated, often under the same term, as for many European countries, or with different names. These include, for instance, the Industrial Internet Consortium in the United States, or 'Made in China 2025' in China, among many further concepts worldwide (Müller & Voigt, 2018).

4.2.2 Small- and medium-sized enterprises

Throughout the European Union, SMEs are defined as enterprises with up to 250 employees, and an annual turnover of up to 50 million Euros. Despite being comparably small in headcount and turnover, SMEs make up about 99 per cent of all enterprises in the European Union. Further, SMEs contribute about 50 per cent of the annual turnover and employ more than 50 per cent of the employees in the European Union (Airaksinen et al., 2015). Comparable characteristics and importance for national economies can be observed worldwide, while the high importance of industrial SMEs is especially observable in the European Union.

Despite their importance for the economy and industrial value creation, SMEs have been regarded considerably less in extant research on Industry 4.0, but are of high importance for the successful implementation of Industry 4.0 across the supply chain (Kagermann et al., 2013; Müller et al., 2018a).

In response to this research gap, the present chapter subsumes research on Industry 4.0 research clusters and compares the most prominent challenges, enablers, and potentials of the researchers found.

4.3 Method

The present chapter is based on a systematic literature review of Industry 4.0 related research in the field of SMEs. Systematic literature reviews allow to generate replicable results, representing a transparent process to minimize the bias and errors of literature research. Also, systematic literature reviews are able to classify and structure extant research, identifying gaps, and based on those findings, presenting recommendations for future research (Tranfield, Denyer, & Smart, 2003).

The systematic literature review was conducted using Scopus, encompassing a nearly complete database for peer-reviewed journal articles. The research strong 'Industry 4.0 AND small- and medium-sized enterprises OR SMEs' in Abstract, Title, and Keywords led to 240 articles found on the database (conducted in September 2019). Limiting these 240 articles to peer-reviewed journal articles (listed in the Scimago Journal and Country Rank) in English language led to 86 articles on Scopus. As several journals relating to purely technology-oriented disciplines (such as 'Manufacturing Letters'), as well as articles focusing on technological solutions being excluded, a sample of 32 journal articles was further examined from Scopus.

Of these 32 articles, several research clusters could be identified that stem from the same group of authors. Only authors that contributed at least two articles were considered for this process, leading to a reduction to 15 articles, stemming from the following research clusters (only naming the author with most articles in the following, or the first named author when having authored the same number of articles): Müller, J. M. (six articles), Rauch, E. (three articles), as well as Mittal and Moeuf (two articles each). Therefore, almost half of the 32 articles found on Scopus can be attributed to four research clusters, of which four are mainly active in Europe.

Following this approach, the five research clusters were extended by a search on Google Scholar for relevant journal articles and conference papers with accessible full text. In particular, articles were searched for that might contain important insights on SMEs in regard to Industry 4.0, but the articles might not have been tagged with the respective keywords or information in their titles and abstracts. Therefore, the final set of papers examined is extended from 12 to 14 papers.

In the following section, the results of this research process are presented.

4.4 Results

Table 4.1 shows the research clusters identified that are prominent in Industry 4.0 related research on SMEs. The research clusters have been identified by respective frequency of the respective researchers, although other research institutions might be involved in single articles. Further, the articles are sorted by their respective occurrence in literature, starting from past to present.

Most occurring publication outlets within the articles found are the *International Journal of Production Research* with three articles. Table 4.2 presents a short description of the articles, including the respective method, main focus of the article, and where it was published.

Interestingly, the majority of research publications is based on qualitative expert interviews and case studies stemming from single countries or literature reviews. This is not surprising for an early state of research and emerging research field, represented by Industry 4.0 and SMEs. Still, future research should attempt to quantify, compare, and generalize extant findings both across industry sectors and across several countries (Müller et al., 2019b; Veile et al., 2019b). Further, the now extended state of

Table 4.1 Research clusters of Industry 4.0 related research on SMEs

Research cluster	Research works	Contributors
Nürnberg (Germany), Salzburg (Austria)	Müller and Voigt (2016) Müller et al., (2017) Müller et al. (2018a) Müller, Kiel, and Voigt (2018b) Müller and Däschle (2018) Müller and Voigt (2018) Müller (2019a)	Müller (7), Voigt (5), Veile (2), Buliga (1), Däschle (1), Kiel (1), Maier (1)
Bozen (Italy)	Seidenstricker, Rauch, and Dallasega (2017) Gualtieri et al. (2018) Orzes et al. (2018)	Rauch (3), Matt (2), Bednar (1), Carabin (1), Dallasega (1), Gualtieri (1), Orzes (1), Palomba (1), Poklemba (1) Rojas (1), Seidenstricker (1), Vidoni (1)
Morgantown (WV, United States)	Mittal et al. (2018) Mittal et al. (2019)	Mittal (2), Khan (2), Romero (2), Wuest (2), Menon (1), Purohit (1)
Paris (France)	Moeuf et al. (2018) Moeuf et al. (2019)	Moeuf (2), Lamouri (2), Pellerin (2), Tamayo-Giraldo (2), Eburdy (1) Tobon-Valencia (1)

research on Industry 4.0, compared to two to three years ago, must be better implemented within the current state of research on SMEs. Also, only few articles take a more holistic view on Industry 4.0 in SMEs, integrating technical, economical, ecological, and social aspects.

From a temporal perspective, while the concept of Industry 4.0 has been introduced in 2011 (Kagermann et al., 2013), Müller and Voigt (2016) were the first authors to address the concept of Industry 4.0 in 2016. Several publications can be observed by different research clusters since 2017, indicating a potential increased research interest in this field for the future.

4.5 Conclusion

For research in the field of Industry 4.0, the present chapter gives an overview on extant research clusters on Industry 4.0 and SMEs, presenting about half of extant research in this area that derives from four research clusters. While three research clusters are situated in Europe, one is found in the United States. However, the latter is composed of an author that

Table 4.2 Description of research articles ($n = 14$)

Authors	Method	Main focus	Publication
Gualtieri et al. (2018)	Case study	Engineering Education and Employees Training required for SMEs in Industry 4.0 experiences from a Smart Factory lab	*IEEE International Conference on Industrial Engineering and Engineering Management (IEEM)*
Mittal et al. (2018)	Literature review	Analysis of smart manufacturing and Industry 4.0 maturity models and their applicability for SMEs	*Journal of Manufacturing Systems*
Mittal et al. (2019)	Literature review	Development of a smart manufacturing adoption framework for the special requirements of SMEs	*International Journal of Production Research*
Moeuf et al. (2017)	Literature review	Development of a framework for SMEs regarding Industry 4.0, particularly for requirements, objectives and technologies.	*International Journal of Production Research*
Moeuf et al. (2019)	Delphi study with 12 experts	Identification and evaluation of critical success factors, risks, and opportunities of Industry 4.0 in SMEs	*International Journal of Production Research*
Müller and Voigt (2016)	Expert interviews with 177 German SMEs	Risks, doubts, and challenges associated with Industry 4.0 from the perspective of SMEs	*Productivity Management*
Müller et al. (2017)	Expert interviews with 68 German SMEs	Cooperation strategies among SMEs and challenges associated that hamper cooperation	*Proceedings of the Hamburg International Conference of Logistics*
Müller et al. (2018a)	Expert interviews with 68 German SMEs	Business model innovation in SMEs through Industry 4.0, associated enablers and barriers, and a stage-gate model for SMEs in Industry 4.0	*Technological Forecasting and Social Change*

Table 4.2 (Continued) Description of research articles ($n = 14$)

Authors	Method	Main focus	Publication
Müller et al. (2018b)	Quantitative questionnaire from 746 German industrial enterprises	Positive and negative influences on Industry 4.0 adoption for economic, ecological, and social aspects, differentiated for SMEs and large enterprises	*Sustainability*
Müller and Däschle (2018)	Quantitative questionnaire from 111 German SMEs	Business model innovation in SMEs through Industry 4.0 and which targets in process innovation of their customers it has	*Processes*
Müller and Voigt (2018)	Quantitative questionnaire from 222 German and 107 Chinese SMEs	Benefits and challenges of Industry 4.0 and Made in China from an economic, ecological, and social perspective and a general comparison of both concepts for SMEs	*International Journal of Precision Engineering and Manufacturing-Green Technology*
Müller (2019a)	Expert interviews with 43 German SMEs	Business model innovation in SMEs through Industry 4.0 within the Business Model Canvas, differentiated for Industry 4.0 providers and users	*Journal of Manufacturing Technology Management*
Orzes et al. (2018)	Focus group study	Implementation barriers to SMEs regarding Industry 4.0	*IEEE International Conference on Industrial Engineering and Engineering Management (IEEM)*
Seidenstricker et al. (2017)	Theoretical development and case study	A methodology for business model innovation through Industry 4.0 in SMEs using the TRIZ nine-field matrix	*Zeitschrift für Wirtschaftlichen Fabrikbetrieb*

comes from Germany. In sum, it can therefore be subsumed that Industry 4.0 related research on SMEs is particularly found in Europe, possibly relating to the emergence of the concept in Germany and the high importance of SMEs for industrial value creation in Europe.

Further, the development of Industry 4.0 related research on SMEs over time becomes visible, highlighting research methods and research focuses used. In this regard, the tendency toward case studies, expert interviews, or literature reviews is apparent, while quantitative approaches with a broader scope or higher generalizability are still scarce. The three main research areas include risks, barriers, or general implementation of Industry 4.0 in SMEs and related frameworks developed, business models and business model innovation, and environmental and social aspects regarding accompanying the economic dimension of Industry 4.0.

The main limitation of this chapter is related to a particular choice in literature selection: The focus in existing research clusters on SMEs. While this approach makes results more comparable and highlights topics from several papers written by the same group of authors, authors of single papers on Industry 4.0 related aspects, and SMEs that might work on future contributions that have not been published yet cannot be regarded in this chapter.

For future research, research should better investigate existing research on SMEs, digitization, and automation, and condense these findings with the state of research on Industry 4.0 and SMEs. The convergence of empirical findings, theoretical models developed, and insights from specific case studies should be integrated in future studies. Further, generating more generalizable results or such that have a broader scope, e.g., across countries or industries, should be considered in order to drive this research stream further.

References

Airaksinen, A., Luomaranta, H., Alajääskö, P., & Roodhuijzen, A. (2015). Statistics on Small and Medium-Sized Enterprises: Dependent and Independent SMEs and Large Enterprises. Retrieved on January 10th 2017 from http://ec.europa.eu/eurostat/statistics-explained/index.php/Statistics_on_small_and_medium-sized_enterprises

Birkel, H. S., Veile, J. W., Müller, J. M., Hartmann, E., & Voigt, K.-I. (2019). Development of a risk framework for Industry 4.0 in the context of sustainability for established manufacturers. Sustainability, 11(2), 384.

Gualtieri, L., Rojas, R., Carabin, G., Palomba, I., Rauch, E., Vidoni, R., & Matt, D. T. (2018). Advanced automation for SMEs in the I4. 0 revolution: Engineering education and employees training in the smart mini factory laboratory. In 2018 IEEE International Conference on Industrial Engineering and Engineering Management (IEEM) (pp. 1111–1115).

Kagermann, H., Helbig, J., Hellinger, A., & Wahlster, W. (2013). Recommendations for implementing the strategic initiative INDUSTRIE 4.0: Securing the future of German manufacturing industry. Final Report of the Industrie 4.0 Working Group, Forschungsunion.

Kiel, D., Müller, J. M., Arnold, C., & Voigt, K.-I. (2017). Sustainable industrial value creation: Benefits and challenges of industry 4.0. International Journal of Innovation Management, 21(08).doi: 10.1142/S1363919617400151.

Mittal, S., Khan, M. A., Purohit, J. K., Menon, K., Romero, D., & Wuest, T. (2019). A smart manufacturing adoption framework for SMEs. International Journal of Production Research, 58(5), 1–19.

Mittal, S., Khan, M. A., Romero, D., & Wuest, T. (2018). A critical review of smart manufacturing & industry 4.0 maturity models: Implications for small and medium-sized enterprises (SMEs). Journal of Manufacturing Systems, 49, 194–214.

Moeuf, A., Lamouri, S., Pellerin, R., Tamayo-Giraldo, S., Tobon-Valencia, E., & Eburdy, R. (2019). Identification of critical success factors, risks and opportunities of Industry 4.0 in SMEs. International Journal of Production Research, 1–17. doi: 10.1080/00207543.2019.1636323

Moeuf, A., Pellerin, R., Lamouri, S., Tamayo-Giraldo, S., & Barbaray, R. (2018). The industrial management of SMEs in the era of Industry 4.0. International Journal of Production Research, 56(3), 1118–1136.

Müller, J. M. (2019a). Business model innovation in small- and medium-sized enterprises: Strategies for industry 4.0 providers and users. Journal of Manufacturing Technology Management, 30(8), 1127–1142.

Müller, J. M. (2019b). Antecedents to digital platform usage in industry 4.0 by established manufacturers. Sustainability, 11(4), 1121.

Müller, J. M., Buliga, O., & Voigt, K.-I. (2018a). Fortune favors the prepared: How SMEs approach business model innovations in industry 4.0. Technological Forecasting and Social Change, 132, 2–17.

Müller, J. M., & Däschle, S. (2018). Business model innovation of industry 4.0 solution providers towards customer process innovation. Processes, 6(12), 260.

Müller, J. M., Kiel, D., & Voigt, K.-I. (2018b). What drives the implementation of industry 4.0? The role of opportunities and challenges in the context of sustainability. Sustainability, 10(1), 247.

Müller, J. M., Maier, L., Veile, J., & Voigt, K.-I. (2017). Cooperation strategies among SMEs for implementing industry 4.0. In Proceedings of the Hamburg International Conference of Logistics (HICL), Vol. 23 (pp. 301–318). doi: 10.15480/882.1462.

Müller, J. M., & Voigt, K.-I. (2016). Industrie 4.0 für kleine und mittlere unternehmen. Productivity Management, 3, 28–30.

Müller, J. M., & Voigt, K.-I. (2018). Sustainable industrial value creation in SMEs: A comparison between industry 4.0 and made in China 2025. International Journal of Precision Engineering and Manufacturing-Green Technology, 5(5), 659–670.

Orzes, G., Rauch, E., Bednar, S., & Poklemba, R. (2018). Industry 4.0 implementation barriers in small and medium sized enterprises: A focus group study. In 2018 IEEE International Conference on Industrial Engineering and Engineering Management (IEEM) (pp. 1348–1352).

Seidenstricker, S., Rauch, E., & Dallasega, P. (2017). Industrie-4.0-Geschäftsmodellinnovation für KMU. ZWF Zeitschrift für wirtschaftlichen Fabrikbetrieb, 112(9), 616–620.

Tranfield, D., Denyer, D., & Smart, P. (2003). Towards a methodology for developing evidence-informed management knowledge by means of systematic review. British Journal of Management, 14(3), 207–222.

Veile, J. W., Kiel, D., Müller, J. M., & Voigt, K.-I. (2019a). Lessons learned from industry 4.0 implementation in the German manufacturing industry. Journal of Manufacturing Technology Management, In press.

Veile, J. W., Schmidt, M.-C., Müller, J. M., & Voigt, K.-I. (2019b). Buyer-supplier relationships in Industry 4.0 – A comparison across industries. European Operations Management Association (EurOMA) Conference 2019, Helsinki.

chapter five

The role of Industry 4.0 on the association between customers' and suppliers' involvement and performance improvement

Guilherme Luz Tortorella
Universidade Federal de Santa Catarina, Florianópolis, Brazil

Alejandro Mac Cawley Vergara
Pontifícia Universidad Catolica, Santiago, Chile

Rogério Miorando
Universidade Federal de Santa Catarina, Florianópolis, Brazil

Rapinder Sawhney
University of Tennessee, Knoxville, Tennessee, USA

Contents

5.1 Introduction

Companies that present the highest likelihood of success are the ones that effectively link their internal value stream with their supply chain (Frohlich and Westbrook, 2001; Krajewski et al., 2015). The benefit obtained from linking internal processes with customer and suppliers reinforces the importance of a proper integration with suppliers and customers as a central aspect to enhance competitiveness beyond usual company's limits (Flynn et al., 2010; Frazzon et al., 2015). In this sense, the establishment of an efficient and responsive supply chain management (SCM), which embraces the flow of value from very beginning at the supplier until the final consumer (Power, 2005), has motivated strategic discussion of several companies (McCormack et al., 2008).

The need for systemic improvement of SCM has generated the need for developing close relationships with both suppliers and customers (Tortorella et al., 2017b), so that waste reduction and quality enhancement is allowed (Marodin et al., 2019). Such improvement approach applied at a supply chain level can be conceptualized as a group of firms whose products, services, and information flow collaboratively and efficiently upstream and downstream in order to reduce wasteful activities and costs, while delivering the exact customer's individual needs (Vitasek et al., 2005; Anand and Kodali, 2008). Many studies (e.g., Cagliano et al., 2006; Lewis, 2006; Blanchard, 2010; Tortorella et al., 2017a) have reported a positive interaction between the adoption of practices for customers' and suppliers' involvement and company's performance improvement.

From a more recent perspective, the advent of Industry 4.0 (I4.0) and its inherent technologies have raised attention for the potential of achieving different levels of operational performance (Landscheidt and Kans, 2016; Tortorella and Fettermann, 2018). I4.0 comprehend features such as interconnected equipment, intelligent products, and interrelated systems. These technologies are systemically joined to establish intelligent production units that control material flow and machines (Kagermann et al., 2013; Fatorachian and Kazemi, 2018). This fact is specifically noticed when considering industries, such as automotive and chemical process, that are undergoing a significant transformation resultant from I4.0 (Lasi et al., 2014).

At first sight, one might expect that concurrently enhancing customers' and suppliers' involvement along with adopting I4.0 technologies would be synergistic. The first approach, as it supports the change of mindset of organizations within a certain supply chain (Naim and Gosling, 2011); and the second one, as it gives a higher level of flexibility and automation in production system, supporting new manufacturing philosophies (Ashton, 2009). Nevertheless, the incorporation of I4.0 into existing management approaches and which processes it can positively underpin still needs further investigation (Kolberg et al., 2016; Tortorella et al., 2018a). Specifically within the SCM and the intrinsic involvement of customers and suppliers, it is noteworthy that the scarcity of research empirically examines the impact of I4.0 to provide higher-performance levels (Gjeldum et al., 2016; Landscheidt and Kans, 2016). Further, as most improvement initiatives are primarily addressed to company's internal processes (Marodin et al., 2017), working on enhancing customers' and suppliers' relationships are usually left to a second round of continuous improvement. Therefore, the incipient understanding of the impact of customers' and suppliers' involvement added by the adoption of disruptive I4.0 technologies entail a differentiated context for improving performance.

Based on this discussion, we raise the following research question: "How do I4.0 technologies moderate the effect of customers' and suppliers' involvement on performance improvement?" Hence, our research examines the moderating effect of I4.0 on the relationship between practices related to customers' and suppliers' involvement and performance improvement in the Brazilian industry. To achieve our objective, we applied a survey to the Brazilian manufacturing companies and captured four SCM performance indicators and 11 practices that aim to enhance involvement of both customers and suppliers (Shah and Ward, 2007). To determine the adoption of I4.0, we asked for the adoption level of digital technologies related to manufacturing processes and product development and service innovation, which are the most suitable ones for the Brazilian industrial scenario (Brazilian National Confederation of Industry, 2016; Tortorella and Fettermann, 2018). Using multivariate data analysis, we expect to shed light on the moderating effect of I4.0 technologies between practices related to customers' and suppliers' involvement and supply chain performance. As previously done by Shah and Ward (2003, 2007), we extend the notion of "bundles" from human resource practices, suggested by Osterman (1994) and MacDuffie (1995), to include the implementation of I4.0 technologies. For these technologies, we defined the following bundles: (i) manufacturing process and (ii) product development/service innovation, as suggested by Tortorella et al. (2018a). For the case of customer- and supplier-related ones, we defined two constructs: (i) involved suppliers and (ii) involved customers. Overall, this chapter expands upon Tortorella et al. (2018b).

Besides its implications to theory, our study allows managers to better understand the advantages and drawbacks of adopting I4.0 into SCM to achieve increased supply chain performance standards. Furthermore, the clear identification of the relationship between these approaches underpins addressing countermeasures that can mitigate eventual barriers, while setting the right expectations along their adoption within the supply chain. Such outcome is especially important in emerging economies' contexts, such as Brazil, where supply chain agents usually present a poor collaborative relationship and the integration among customers and suppliers are either mislead or subjected to the influence of companies with higher bargain power (McCormack et al., 2008; Marodin et al., 2016, 2017).

5.2 *Literature review and hypotheses*

5.2.1 *Industry 4.0*

The increasing implementation of automation is argued to enhance product quality and manufacturing efficiency (Kagermann et al., 2013). The term "Industrie 4.0" was publicly acknowledged in 2011, when representatives from business, politics, and academia highlighted this concept to reinforce German manufacturers' competitiveness. I4.0 aims at describing a dynamic production that incorporates Information and Communication Technologies (ICT) to facilitate a highly customized manufacturing. Analogous to manufacturing, I4.0 can significantly affect supply chains, business models, and services (Schmidt et al., 2015; Dalenogare et al., 2018; Queiroz and Telles, 2018).

Although it's widely acknowledged, several companies still struggle with I4.0 concepts (Sanders et al., 2016; Tortorella and Fettermann, 2018). Hence, to drive I4.0 adoption and reduce misconceptions, some guidelines have been proposed (Qin et al., 2016; Bibby and Dehe, 2018). Lee et al. (2015) suggest a five-level model that guides I4.0 deployment for manufacturing context. The German Government used a 62-item and nine-dimensional model for assessing I4.0 (Schumacher et al., 2016). Complementarily, Anderl (2014) proposes the utilization of a practical implementation roadmap, framed on six dimensions and five maturity levels. Ganzarain and Errasti (2016) present a three-stage I4.0 model with specific application for small and medium enterprises.

Further, many government institutions have assessed the I4.0 maturity level of the respective companies. In developing economies, Brazilian National Confederation of Industry (2016), for example, surveyed more than 2,000 manufacturers to identify the main challenges for I4.0 adoption. The Indian government developed an initiative that aims at positioning India as a hub for I4.0 (Forbes India, 2016). In Mexico, government has developed an I4.0 roadmap, due to the significant level of high-tech products exported (Mexican Ministry of Economy, 2016). Nevertheless, many further opportunities are still open to understand the benefits of I4.0 implementation in developing economies.

There are a number of technologies that enable I4.0. But it is the connection between the physical and the virtual environments what has become a relevant contribution of I4.0 (Hermann et al., 2016). In this sense, the cyber-physical systems (CPS) allow managing interconnected systems with respect to their physical assets and computational capabilities (Baheti and Gill, 2011). To achieve that, novel technologies are necessary to allow exchanging information about industrial processes in a timely, reliable, and possibly uniform way (Wollschlaeger et al., 2017). At a process level, some of the technologies that enable a timely and reliable flow of information are digital sensing, Supervisory Control and Data Acquisition (SCADA) (Karnouskos et al., 2014), flexible production systems (Weyer et al., 2015), and integrated engineering and manufacturing. From a product development and service perspective, Rüßmann et al. (2015) indicate that the use of simulation, horizontal and vertical systems integration, Internet of Things (IoT), cybersecurity, and cloud technologies are examples of technologies that can transform industrial production.

The concurrent implementation of I4.0 technologies with other managerial approaches is not a new subject in the literature. Tortorella and Fettermann (2018) studied 41 measures related to lean production (LP) and 10 digital automation technologies. In their study, they consolidated all corresponding measures as single dimensions, disregarding associations between specific sets of practices and technologies. Similarly, Buer et al. (2018), based upon a literature review, highlighted the existing gaps that still need to be examined in the relationship between LP and I4.0, while Sanders et al. (2016, 2017) have envisioned how this relationship occurs at a practice-technology level. Overall, none of these studies specifically address the moderating effect of I4.0 on the relationship between customers/suppliers-related practices and supply chain performance, motivating our research.

5.2.2 *Lean supply chain management*

The establishment of a Lean Supply Chain Management (LSCM) shifts from a traditional approach, where short-term profit goals are mainly dependent upon price negotiation, to a strategic commitment with supply chain partners, so that waste is systematically and cooperatively reduced (Yusuf et al., 2004; Agarwal et al., 2006; Marodin et al., 2017). LSCM states that production must follow customers' needs under the logic of a pull system of goods (Morris et al., 2004; Jasti and Kodali, 2016). Hence, material and information flows are accounted as part of the LSCM, encouraging a collaborative relationship among partners that generate operational advantages (Cox et al., 2007; Tortorella et al., 2017a).

Various companies still have difficulties with LSCM implementation because of the little awareness and inappropriate methodologies (Marodin et al., 2017, 2019). Indeed, lean adaptation to SCM is quite difficult (Hines et al., 2004) due to a few aspects, such as (i) identification and quantification of waste is easier in internal productive operations rather than supply chain and (ii) senior managers are more likely to control manufacturing processes than supply chain, as it demands the involvement of the entire chain partners (Anand and Kodali, 2008; Soni and Kodali, 2013). Additionally, various studies focused on specific elements of LSCM, while very few approached the relationships with both suppliers (upstream) and customers (downstream) (Anand and Kodali, 2008; Jasti and Kodali, 2015; Riet et al., 2015; Tortorella et al., 2017a). In fact, Moyano-Fuentes et al. (2019) argued that LSCM may be composed of two dimensions: one related to the operational aspects of lean supply chain strategy and the second related to lean supply chain planning. The active involvement of customers and suppliers would be more associated with the latter dimension.

Overall, implementing lean across a multilevel supply chain is an additional challenge (Bruce et al., 2004; Taylor, 2006). In this sense, Pérez et al. (2010) specified some challenges that need to be overcome so that a true LSCM is achieved. Particularly for this research, we investigate the impact of the concurrent implementation of practices that foster customers' and suppliers' involvement and I4.0 technologies adoption on performance improvement.

5.2.3 Suppliers' involvement and I4.0

From a broad perspective, LP implementation seeks to eliminate waste and process variability, maximizing value addition to customers while improving operational performance (Shah and Ward, 2003; Stentoft and Vagn, 2013). As LP understanding has evolved from a practice-oriented perspective to a value system approach across multiple organizations (Hines et al., 2004), the adaptation of the accrued knowledge expanded the usual organizational boundaries. Recent studies (e.g., Moyano-Fuentes and Sacristán-Díaz, 2012; Bhamu and Singh Sangwan, 2014; Jasti and Kodali, 2015) highlight that the value chain is deemed as the new frontier of LP research.

In terms of technologies, existing developments have been motivated by the need for higher efficiencies in manufacturing processes. However, those efforts have been mainly focused on advances of the focal company rather than on the whole supply chain (Tortorella et al., 2018b). Thus, these technologies have led to significant but isolated efficiency and quality improvements in processes and products (Schumacher et al., 2016). With I4.0, suppliers may be able to provide an integrated technological platform into customers' products through the proper adjustment of the necessary interfaces (Knoppen et al., 2015). Such platform not only facilitates software integration, but also bundles competences to meet customers' expectations (Kampker et al., 2016). Further, inappropriate information flow between manufacturers and suppliers might entail an additional waste. Suppliers need to be constantly informed about the status of the products and services provided by them to address an adequate and fast countermeasure to settle discrepancies (Sanders et al., 2016). Therefore, I4.0 bears this interaction by providing the necessary technologies, such as IoT and Big Data, to obtain and provide immediate suppliers' feedback, overcoming traditional bureaucracies and slow and nonassertive communication channels (Dworschak and Zaiser, 2014).

Additionally, a LSCM stands for a timely delivery, which is usually an issue for most companies due to a diversity of reasons such as incomplete and mismatched shipments of goods, and unexpected time delays during delivery (Flynn et al., 2010; Dora et al., 2016). IoT adoption presupposes the utilization of ICT devices, which can handle information about goods and track their status until their final destination (Ashton, 2009; Shariatzadeh et al., 2016). This kind of technology might support not only timely delivery, but also generate the opportunity for delivery routes optimization. Thus, based on these arguments, we formulate the following hypothesis to investigate the effect of I4.0 on the relationship between lean practices related to suppliers and performance improvement:

> *H1: The adoption of I4.0 technologies positively moderates the effect of Suppliers' Involvement on performance improvement.*

5.2.4 *Customers' involvement and I4.0*

The constant involvement of customers allows companies to better predict and understand their needs (Shah and Ward, 2007; Tortorella et al., 2017a). Such involvement can occur during several phases of the productive flow. For instance, the identification of customer's perceived value to generate new and profitable products and services (Zacharia et al., 2019), and exploring synergies between processes, people, and technologies has been denoted as lean product development (LPD) (Tortorella et al., 2016). Jasti and Kodali (2015) emphasized the importance of involved customers to fully achieve a LSCM by defining a bundle of practices named "Customer relationship management". As successful organizations are based on how well they deliver products and services to customers, they should similarly value both internal and external customers. In this sense, organizations should invest efforts to maintain long-term relationships with customers and analyze their needs to provide better services in terms of cost and quality (Soni and Kodali, 2013).

With regards to I4.0, Kampker et al. (2016) mentioned that as novel technologies are introduced, a change in customers' expectation can be noticed. In this sense, the possibilities raised by technological advances in terms of communication, virtual reality, and connected systems are extremely relevant to accelerate time-to-market and avoid increased costs for product development (Rauch et al., 2016). The existing tools (e.g., Quality Function Deployment – QFD) are usually limited to the number of customers' requirements and relationships with product design (Sanders et al., 2016). In turn, Big Data, for instance, seems to enable complex calculations and processing of large volume of data (Li et al., 2015), allowing manufacturers to differentiate categories of customers and providing more assertive products and solutions (Shrouf et al., 2014). Thus, it is reasonable to expect that the application of I4.0 technologies together with customer-related lean practices opens new ways to manage customers' information. We argue that, if developed properly under the lean principles, I4.0 technologies can support customer-related practices and increase its operational benefits in the supply chain. So, we propose the following hypothesis with regard to customer involvement and I4.0:

> *H2: The adoption of I4.0 technologies positively moderates the effect of customers' involvement on supply chain performance improvement.*

The hypotheses indicate the moderating role played by I4.0 incorporation into the relationship between customers' and supplier's involvement and performance improvement. Thus, the conceptual model investigated in this research is illustrated in Figure 5.1.

5.3 Method

5.3.1 Sample selection and characteristics

National culture is claimed to impact LP practices implementation (Kull et al., 2014). As this research focused on Brazilian industrial context, the surveyed sample was limited to leaders from these companies.

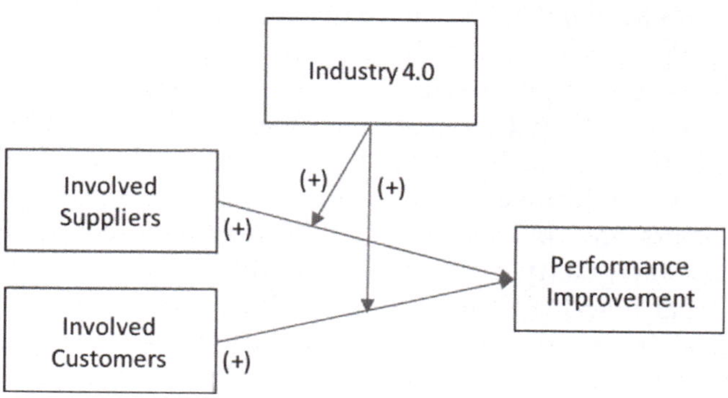

Figure 5.1 Investigated model.

Furthermore, respondents should be experienced in LP and I4.0 technologies, as suggested in studies with similar subjects, such as Tortorella and Fettermann (2018) and Tortorella et al. (2018b). As respondents that meet both criteria are extremely scarce in this industrial context, the studied sample comprised companies from various industrial sectors. Moreover, as the actual extent of LP amongst different sectors is still poorly understood (Tortorella et al., 2015), the utilization of a cross-industry sample is also recommended to provide more robust analysis. Previous studies (e.g., Shah and Ward, 2003, 2007; Marodin et al., 2016; Tortorella et al., 2016) that carried out surveys on LP have also used a similar strategy.

The questionnaire consisted of four parts. The first part evaluated the perceived improvement levels over the past three years of company's supply chain performance based upon the following aspects: (i) productivity, (ii) delivery service level, (iii) inventory level, and (iv) quality. We used a Likert scale, that ranged from 1 (worsened significantly) to 5 (improved significantly), to assess that. The second part collected information on respondents' companies and supply chain. The third part evaluated the implementation level of the supplier- and customer-related lean practices (Shah and Ward, 2007), which comprised 11 items. These practices were extensively referenced in several studies in supply chain (e.g., Sousa and Voss, 2008; Bozarth et al., 2009; Zhao et al., 2011), justifying their application in our study. Practices' statements were assessed in a Likert scale varying from 1 (fully disagree) to 5 (fully agree). Finally, the last part examined the adoption level of I4.0 technologies within the studied sample. For that, 10 digital technologies were listed according to the Brazilian National Confederation of Industry (2016). These technologies are argued to be the most suitable ones in the Brazilian industrial context, and have also been used as a proxy for I4.0 implementation by Tortorella and Fettermann (2018). Similarly, a 5-point Likert scale, in which 1 referred to "not used" and 5 indicated "fully adopted", was applied to measure these technologies.

As there is a restricted number of Brazilian companies that have mutually implemented customer- and supplier-related practices into their SCM and initiated the adoption of I4.0 technologies, collecting data was particularly difficult (Marodin et al., 2016; Tortorella et al., 2018b). Therefore, the questionnaire was applied with 147 leaders from manufacturing companies, who have attended LP executive education courses in a large Brazilian University four times during 2017. A first data collection occurred in February 2017 with students from the first class, followed by other three opportunities with students from classes that happened in April, July, and September of the same year. The sample characteristics are displayed in Table 5.1. Respondents were mostly from large-sized firms (55.1%) and 49.6% belonged to metal-mechanic sector. The majority of them (65.9%) were from the first and second tiers and,

Table 5.1 Sample characteristics ($n = 147$)

Category	Description	Quantity	Percentage (%)
Tier level	1 and 2	97	65.9
	3 and 4	50	34.1
Company size	Large (≥500 employees)	81	55.1
	Small and medium (<500 employees)	66	44.9
Technologic intensity	High and medium-high	79	53.7
	Low and medium-low	68	45.3
Industrial sector	Metal mechanic	73	49.6
	Chemical	19	12.9
	Food	13	8.8
	Textile	7	4.8
	Others	35	23.8

regarding companies' technological intensity, 53.7% were from companies categorized as high or medium-high, according to Brazilian National Confederation of Industry's (2016) classification. The whole sample comprised companies located in South (70%) and Southeast (30%) of Brazil, which are the most industrialized regions in Brazil (IBGE, 2012).

5.3.2 *Sample and method bias*

For nonresponse bias test, we followed the procedure recommended by Armstrong and Overton's (1977), which checks for equality of variances and means applying Levene's test and *t*-test, respectively, among respondents of each class. No statistical differences in means and variance were identified among early (first and second classes) and late (third and fourth classes) respondents (p-value < 5%).

Using single respondents to answer both independent and dependent variables is claimed to enhance common method bias. Hence, we applied several techniques, indicated by Podsakoff and Organ (1986) and Podsakoff et al. (2003) to curb these effects. First, in the questionnaire design, we placed the dependent variables (performance indicators) far from the independent ones (customer- and supplier-related practices, and I4.0 technologies). Additionally, we included verbal labels to scales of each questionnaire, as previously performed by studies with similar single respondents' issues (e.g., Marodin et al., 2019; Tortorella and Fettermann, 2018). Further, respondents' anonymity was clearly informed at the beginning of the questionnaire application. It is noteworthy that all respondents played strategic leadership roles in their companies, ensuring the relevance of their responses.

Finally, we also checked for Harman's single-factor test with an exploratory factor analysis (EFA) as a statistical remedy to address potential issues (Malhotra et al., 2006). Results for Harman's test including all variables showed that the first factor embraced 21.94% of the variation, indicating that no single component accounted for most of the variance and issues related to common method variance could be neglected.

5.3.3 Construct validity and reliability

Initially, supported by IBM® SPSS Statistics, we performed an EFA to verify the best combination among the 11 customer- and supplier-related practices included in the questionnaire. Thus, a principal component analysis (PCA) with varimax rotation was carried out to extract orthogonal components (Hair et al., 2014). Factor loadings whose values were larger than 0.50 were considered accepted (Tabachnick and Fidell, 2007). We were able to replicate results through oblique rotation in order to verify orthogonality, whose results indicated that the components were quite similar. Table 5.2 shows that all measures reasonably loaded into two constructs. The first construct, denoted as [INV_SUP], represents practices that foster suppliers' involvement, such as "feedback on quality and delivery performance" and "involvement in new product development process". The second one, [INV_CUST], refers to practices that enhance customers' involvement, such as "involvement of customers in current and future product offerings" and "sharing current and future demand information with marketing department" (Shah and Ward, 2007). Reliability of both constructs was evidenced, since they demonstrated a Cronbach's alpha larger that 0.7 (Meyers et al., 2006).

Analogously, for I4.0 technologies, we performed an EFA to identify technologies that may be grouped because they present much of their variance in common (Hair et al., 2014). In this sense, all 10 technologies were inputted for PCA with varimax rotation to extract orthogonal components. Results for PCA indicated the extraction of two components that could be classified according to their focus (see Table 5.3): (i) process [PROCESS] and (ii) product development and service innovation [PROD_SERV]. The first one, [PROCESS], comprises technologies that are mainly focused on enhancing manufacturing processes, such as digital automation and manufacturing execution system. Complementarily, [PROD_SERV] technologies, such as virtual models and IoT, seek to improve and support processes related to product development and services innovation. We also replicated results using oblique rotation so that orthogonality is tested, which resulted in similar components. It is noteworthy that similar results for I4.0 constructs were found by Tortorella et al. (2018a, 2018b), which provide additional empirical evidence to support the establishment of these constructs. Through the determination of Cronbach's

Table 5.2 PCA results for customers' and suppliers' involvement practices

LSC operational measures	Mean values	Std. dev.	Communalities	Factor loadings 1	Factor loadings 2	
We frequently are in close contact with our suppliers	3.53	1.07	0.652	0.797		Involved suppliers
We give our suppliers feedback on quality and delivery performance	3.29	1.08	0.728	0.842		
We strive to establish long-term relationship with our suppliers	3.56	1.03	0.715	0.843		
Suppliers are directly involved in the new product development process	2.81	1.11	0.513	0.667		
Our key suppliers deliver to a plant on just-in-time basis	2.66	1.08	0.477	0.607		[INV_SUP]
We evaluate suppliers on the basis of total cost and not per unit price	2.68	1.11	0.382	0.561		
We frequently are in close contact with our customers	3.61	1.05	0.547		0.552	Involved customers
Our customers give us feedback on quality and delivery performance	3.61	1.07	0.635		0.707	
Our customers are actively involved in current and future product offerings	3.29	1.15	0.806		0.883	
Our customers are directly involved in current and future product offerings	3.16	1.21	0.862		0.923	[INV_CUST]
Our customers frequently share current and future demand information with marketing department	2.95	1.25	0.689		0.809	
			Eigenvalues	5.219	1.787	
			Initial percentage of variance explained	47.443	16.246	
			Rotation sum of squared loadings (total)	3.631	3.375	
			Percent of variance explained	33.011	30.678	
			Cronbach's alpha (sample $n = 147$)	0.846	0.882	
			Kaiser-Meyer-Olkin	0.872		
			Bartlett test of sphericity (χ^2/dF)	881.20/55 (p-value < 0.01)		

Extraction method: Principal component analysis. Rotation method: Varimax with Kaiser normalization.
All empty values indicate factor loadings lower than 0.5.

Table 5.3 PCA results for I4.0 technologies

I4.0 technologies	Mean values	Std. dev.	Communalities	Factor loadings 1	Factor loadings 2	
Digital automation without sensors	2.63	1.22	0.516		0.719	Manufacturing process
Digital automation with process control sensors	2.74	1.28	0.756		0.806	
Remote monitoring and control of production through systems	2.43	1.34	0.722		0.782	
Digital automation for operating conditions identification	2.30	1.30	0.706		0.769	
Integrated engineering systems	2.46	1.27	0.583		0.573	[PROCESS]
3D printing	2.01	1.18	0.539	0.631		Product development and service innovation
Augmented reality	1.94	1.20	0.575	0.749		
Big Data	2.27	1.28	0.637	0.765		
Cloud computing	2.16	1.24	0.669	0.803		
Internet of Things (IoT)	2.10	1.22	0.569	0.629		[PROD_SERV]
Eigenvalues				5.092	1.178	
Initial percentage of variance explained				50.922	11.783	
Rotation sum of squared loadings (total)				3.172	3.098	
Percent of variance explained				31.721	30.984	
Cronbach's alpha (sample $n = 147$)				0.860	0.800	
Kaiser-Meyer-Olkin				0.852		
Bartlett test of sphericity (χ^2/dF)				751.63/45 (p-value < 0.01)		

Extraction method: Principal component analysis. Rotation method: Varimax with Kaiser normalization.

All empty values indicate factor loadings lower than 0.5.

Table 5.4 PCA results for performance improvement (dependent variable)

Performance indicators	Mean values	Std. dev.	Communalities	Factor loadings
Delivery service level	3.61	0.97	0.677	0.823
Quality	3.50	1.03	0.778	0.882
Productivity	3.54	1.09	0.766	0.875
Inventory level	3.16	1.08	0.650	0.806
			Eigenvalues	2.871
			Initial percentage of variance explained	71.785
			Cronbach's alpha (sample $n = 147$)	0.860
			Kaiser-Meyer-Olkin	0.797
			Bartlett test of sphericity (χ^2/dF)	289.86/6 (p-value < 0.01)

Extraction method: Principal component analysis.

alpha, reliability was assessed for each construct (Meyers et al., 2006). All obtained alpha values were larger than 0.80. The mean values of the technologies integrated into each construct weighed by their corresponding factor loadings obtained from the PCA indicated the response value for each construct.

Finally, with respect to the improvement level on performance, a scale was constructed based upon the PCA of the four indicators. Dependent variable was represented by the factor scores. Table 5.4 displays that the four indicators loaded into a single component, whose eigenvalue was 2.871 and represented almost 72% of the variation, and Cronbach's alpha value of 0.860. Hence, all performance indicators were grouped into one dimension denoted as [PERFORMANCE].

Finally, we have standardized the values for focal predictors, moderators, and dependent variable to address possible multicollinearity, as also performed in Marodin et al. (2017). In this sense, the correlation matrix for all variables are shown in Table 5.5. Additionally, as recommended by Hair et al. (2014), we tested for normality, linearity, and homoscedasticity between independent and dependent variables to confirm the demanded conditions for an ordinary least square (OLS) regression analysis. We used residuals to verify normality of the error term distribution. Linearity was checked with plots of partial regression for each model, whose results suggested that we could not reject the hypothesis of adherence to the normal distribution of residuals for any model (Kolmogorov-Smirnov test p-value < 0.05). We assessed homoscedasticity using plots of standardized residuals against predicted value and visual examination. All tests confirmed the required assumptions.

Table 5.5 Correlation matrix for all independent, dependent, potential moderating, and control variables

Variables	1	2	3	4	5	6	7	8
1. Size	-	0.088	−0.054	0.203*	0.102	0.235**	0.243**	0.131
2. Tier		-	−0.014	0.044	−0.081	0.041	0.081	0.151
3. Technological intensity			-	−0.235**	−0.173*	−0.082	−0.122	−0.168*
4. INV_SUP				-	0.488**	0.457**	0.275**	0.446**
5. INV_CUST					-	0.456**	0.288**	0.347**
6. PROCESS						-	0.615**	0.419**
7. PROD_SERV							-	0.274**
8. PERFORMANCE								-
Cronbach's alpha				0.846	0.882	0.800	0.860	0.860
Composite reliability (CR)				0.859	0.885	0.822	0.871	0.903

* Correlation is significant at the 0.05 level (two-tailed).
** Correlation is significant at the 0.01 level (two-tailed).

5.4 Results and discussions

We carried out three OLS hierarchical linear regression models to examine the proposed hypotheses. Results, depicted in Table 5.6, show unstandardized coefficients. As suggested by Goldsby et al. (2013), unstandardized coefficients can reproduce a standardized effect. The first stage (Model 1) of the hierarchical process analyzed only the effect of control variables (company size, tier level, and technological intensity) on performance improvement. Then, in Model 2, the direct effects of [INV_SUP], [INV_CUST], [PROCESS], and [PROD_SERV] on the dependent variable were included. In Model 3, the moderating effects (represented by the interaction terms) of [PROCESS] and [PROD_SERV] were finally included. Variance inflation factors (VIF) for all regression models were lower than 4.0, which denotes that multicollinearity issues could be disregarded.

Results indicate that, for performance improvement, both the inclusion of the independent variables and interaction terms showed a significant change in R^2. Therefore, the model with the highest capacity of performance improvement prediction was Model 3, which included the moderating effects (F-value = 6.972; p-value < 0.01) and explained 31.0% of the variance.

It is noteworthy that tier level (control variable) appears to be positively impacting ($\hat{\beta} = 0.534$; p-value < 0.05), which indicates that as companies become farther from the end customer, performance improvement tends to increase. Although it might seem a counterintuitive finding, it is aligned with findings from Tortorella et al. (2017a, 2017b). These studies

Table 5.6 OLS hierarchical regression models[a]

Variables	Performance Model 1	Model 2	Model 3
Size	0.239	−0.003	0.006
Tier	0.460*	0.461*	0.534**
Technological intensity	−0.153*	−0.036	−0.049
[INV_SUP]		0.235***	0.202**
[INV_CUST]		0.178**	0.145*
[PROCESS]		0.220**	0.313***
[PROD_SERV]		0.029	−0.045
[INV_SUP] × [PROCESS]			−0.026
[INV_SUP] × [PROD_SERV]			0.045
[INV_CUST] × [PROCESS]			−0.323***
[INV_CUST] × [PROD_SERV]			0.218**
F-value	3.069**	8.871***	6.972***
R^2	0.060	0.309	0.362
Adj. R^2	0.041	0.274	0.310
Change[b] in R^2		0.248***	0.053**

Notes: n = 147.

[a] Unstandardized regression coefficients are reported. All VIF values are below 4.
[b] Change in R^2 reports results compared with the previous model.
* Coefficient significant at 10%.
** Coefficient significant at 5%.
*** Coefficient significant at 1%.

suggest that companies from tier levels closer to the end consumer (e.g., first and second tiers) often purchase raw material from large-sized companies, such as steel mills and foundries. Particularly in Brazilian industry context, the steel mills sector is claimed to be an oligopoly (Marodin et al., 2016), entailing a higher bargain power for these companies at negotiation and problem-solving activities with other supply chain agents. Therefore, since our study sample is comprised mostly of metal-mechanic companies (49.6%), this contextual effect might be perceived in our results. Additionally, the positive direct effect of [INV_SUP] ($\hat{\beta} = 0.202$; p-value < 0.05) suggests that much emphasis has been applied into improving the relationship with upstream partners of the supply chain, which also corroborates to a better performance from companies from tiers three and four.

When analyzed separately, this model shows that the I4.0 construct of [PROCESS] is positively related to performance improvement ($\hat{\beta} = 0.313$; p-value < 0.01), while [PROD_SERV] does not seem to present a significant direct effect. Koufteros et al. (2005) show a similar result in which supplier integration on product innovation has presented a contradictory impact. For instance, they indicate that assigning suppliers with product development responsibilities may undermine organization's ability to offer new products and features, deteriorating product innovation. However, surprisingly, results indicate that when the interaction terms are taken into consideration, the concurrent adoption of [PROCESS] technologies negatively moderates ($\hat{\beta} = -0.323$; p-value < 0.01) the effects of [INV_CUST]. This was also observed by Ray et al. (2005), who concluded that, when low levels of shared knowledge are observed between companies and their customers, the investment in information technology may even reduce process performance. On the other side, as expected, [PROD_SERV] positively moderates ($\hat{\beta} = 0.218$; p-value < 0.05) the effects of [INV_CUST] on performance improvement. The effects of these interactions are illustrated in Figures 5.2 and 5.3. Figure 5.2 indicates that as the implementation of [INV_CUST] practices becomes more extensive, the adoption of [PROCESS] technologies should be reduced to perceive higher performance improvement levels. In turn, Figure 5.3 shows that for higher levels of [INV_CUST], companies tend to achieve higher improvement levels for performance if I4.0 technologies denoted as [PROD_SERV] are highly adopted. Although the moderating variables (PROCESS and PROD_SERV) have contrary impacts, the change on [INV_CUST] adoption appears to override this interaction indicating a positive change in the supply chain

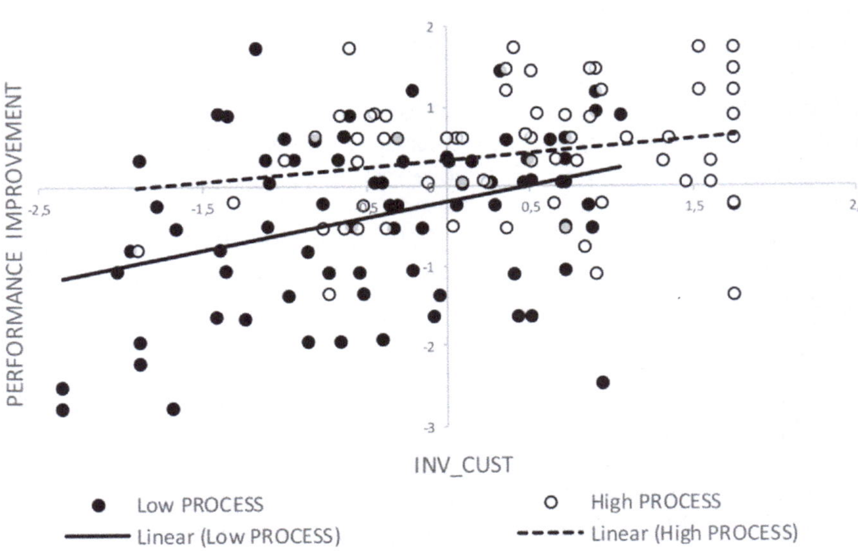

Figure 5.2 Performance improvement and customers' involvement moderated by I4.0 process-related technologies.

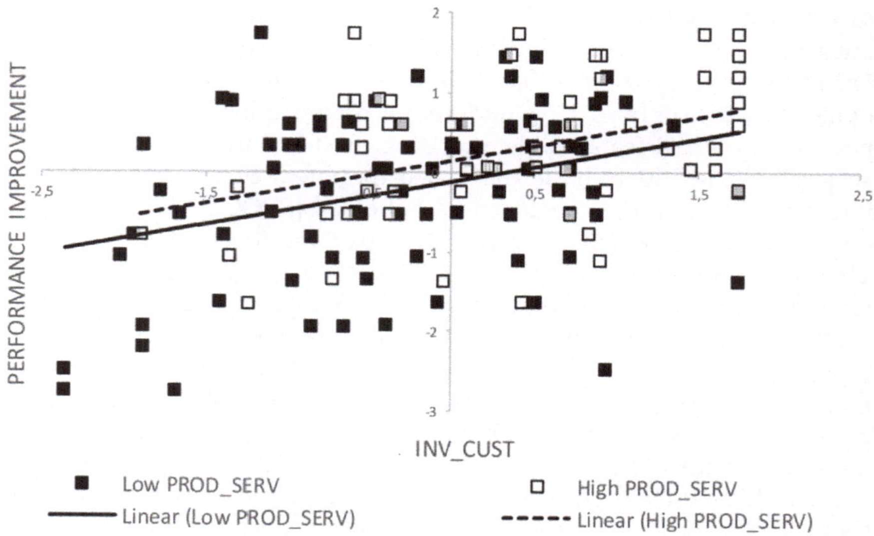

Figure 5.3 Performance improvement and customers' involvement moderated by I4.0 product- and service-related technologies.

performance. This was also noted by Fang (2008), who showed that new product development speed is drastically increased when customer participation and information are used to identify and resolve problems. These findings empirically support that both [PROCESS] and [PROD_SERV] significantly influence the effect of customer relationship practices on performance improvement. Nevertheless, I4.0 technologies related to manufacturing processes were found to be negatively moderating, meanwhile technologies focused on product development and service innovation are positively moderating. Those results are unique outcomes and further discussions need to be addressed.

First, regarding the [PROD_SERV] moderation, it is reasonable to expect that I4.0 technologies that primarily aim at supporting and facilitating product development and service innovation are positively moderating the effect of customer-related practices. In fact, recent research on I4.0 (e.g., Lee et al., 2014; Porter and Heppelmann, 2015) has already pointed such potential benefit. Our results add new insights to this body of knowledge. We argue that the isolated incorporation of novel technologies into ill-structured product development and service innovation processes will not bring the expect results, which is observed by the lack of a significant direct effect of [PROD_SERV]. However, if customers are actively involved and, hence, modifications on organizational mindset with respect to the importance of a healthy collaboration with customers are observed, I4.0 technologies (e.g., 3D printing and cloud services associated with product) can significantly enhance performance of delivery, quality, productivity, and inventory.

Second, the negative moderating impact of [PROCESS] technologies on the association between [INV_CUST] and performance improvement indicates that Brazilian manufacturers are still presenting a disconnected improvement approach. In other words, while our findings show that manufacturing process-related technologies do contribute to performance improvement when separately adopted, the same does not occur when customer-related practices are concomitantly considered. This outcome converges to findings from Marodin et al. (2019) and Tortorella et al. (2017a), who suggested that Brazilian companies usually present their improvement initiatives on manufacturing processes apart from the perspective on the supply chain flow of value. In this sense, our research provide empirical evidence to argue that, particularly for the performance improvement of Brazilian companies, the implementation focus of customer-related practices might be diverging and, thus, competing with the benefits provided by the adoption of [PROCESS] technologies, such as SCADA and machine/process control sensors (Frank et al., 2019).

Overall, obtained results partially confirm *H2*, indicating that I4.0 technologies do moderate the effects of customers' involvement on performance improvement, although the direction of such moderation may depend on the focus of the adopted technologies. In opposition, our findings do not bear any I4.0 moderating effect claimed in hypothesis *H1*, which is related to practices focused on suppliers' involvement.

5.5 Conclusions

This chapter conducted a survey-based study with information from 147 Brazilian manufacturers to investigate the moderating effect of the adoption of I4.0 technologies on the association between customers' and suppliers' involvement practices and performance improvement. Contributions of this study are from both theoretical and practical perspectives, and are described in the following sections.

5.5.1 Theoretical contribution

First, in theoretical terms, we provided empirical evidence to affirm that constructs of I4.0 actually moderate the effects of practices related to customers' and suppliers' involvement, although its extension may not occur always as expected. This fact was especially observed for practices focused on customers' involvement, which highlights the need for reinforcing the integration between supplier-related practices and I4.0 technologies. Further, our study indicates that the single implementation of cutting-edge technologies will not necessarily lead to better performance.

In fact, the application of these technologies over ill-structured organizational processes, either manufacturing or product/service processes, can jeopardize performance, leading to outputs contrary to expectations. In this sense, while I4.0 can fundamentally modify SCM and lead to novel levels of performance, it may also raise the need for a more mature mindset of organizations in order to avoid the illusion of achieving higher performance levels through the simple acquisition of technology and not by actually engaging customers and suppliers in a systematic fashion.

Additionally, our findings indicate that, specifically in the Brazilian context (emerging economy), those I4.0 technologies which are focused on product/services development do not have a direct influence on performance improvement. These results are similar to those obtained by Dora et al. (2016) and Sanders et al. (2016), who suggest that unless supply chain agents are actually working in a collaborative and supportive relationship, the applied practices and/or technologies will not succeed or will present irrelevant marginal gains in performance. Further, our results also show that, when product/services technologies are concurrently implemented along with customers' involvement practices, they positively impact performance improvement.

5.5.2 *Practical implication*

Second, with regards to practical implications, our findings indicate that manufacturing companies in Brazil may still lack a systemic improvement approach that integrates the whole value system, which is claimed to be the new frontier of LP (Hines et al., 2004). In other words, companies' improvement initiatives tend to be myopic, providing changes on manufacturing shop floor processes regardless their effect on supply chain (customers and suppliers), and vice versa. Therefore, our research indicates the need for a holistic approach for implementing I4.0 while actively engaging customers and suppliers. With the support from advanced ICT on manufacturing, product development and service innovation processes along with an enhanced participation and involvement of their customers and suppliers, manufacturing companies can indeed move into new performance standards.

Moreover, although the current maturity of the study sample indicates a myopic implementation of I4.0 and customer- and supplier-related practices, our results already emphasize that manufacturers that highly involve customers and adopt technologies focused on product development and service innovation are more likely to achieve higher improvement in their performance levels. Such fact corroborates with indications from Rauch et al. (2016) and Sanders et al. (2016), providing evidence to managers that both approaches are synergic and unified small efforts on both may lead to significant increments on performance.

5.5.3 Limitations and future research

This study has limitations that should be mentioned. As all the dataset was comprised of Brazilian firms, their perceptions might be related to specific socioeconomic issues. This is an important aspect since lean implementation has been claimed to be highly dependent on sociocultural characteristics. This fact restricts findings to this contextual condition, even though it improves the likelihood that they may be generalizable to those supply chains and to others in similar regions and contexts. However, it is noteworthy that the need for increasing the sample to other developing economies with sociocultural similarities in order to provide more robust findings that could be compared to developed economies' context. Second, data collection was performed utilizing a perceptual instrument, which has a few drawbacks. Despite the fact that adequate remedies have been undertaken to mitigate biases, further research could restrict sample and carry out studies with a larger sample size. As findings are based upon respondents' perceptions, there are specific limitations related to performance improvement. Therefore, further research could provide more insightful findings from gathering actual supply chain data for performance improvement.

Moreover, our study applied an instrument that comprised a set of customer- and supplier-related practices and I4.0 technologies that can be further integrated with complementary practices or technologies. More specifically, practices included in this study approach only the direct interaction between companies and their direct suppliers and customers. In this sense, an overall perspective of the actual supply chain that the company belongs to is much harder to obtain and, hence, must be explored on future research. Regarding I4.0, the technologies included in the questionnaire were obtained from Brazilian National Confederation of Industry (2016), which suggest that these are the most likely to be adopted in Brazil. However, I4.0 adoption implies the utilization of further technologies that were not included in this research. Similarly, future research may integrate a wider set of I4.0 technologies even to allow comparative studies among countries (developed and developing economies).

References

Agarwal, A., Shankar, R., and Tiwari, M. (2006), "Modeling the metrics of lean, agile and leagile supply chain: an ANP-based approach", *European Journal of Operational Research*, Vol. 173, pp. 211–225.

Anand, G., and Kodali, R. (2008), "A conceptual framework for lean supply chain and its implementation", *International Journal of Value Chain Management*, Vol. 2, No. 3, pp. 313–357.

Anderl, R. (2014), "Industrie 4.0: advanced engineering of smart products and smart production", *Proceedings of the 19th International Seminar on High Technology*, Piracicaba, Brazil.

Armstrong, J., and Overton, S. (1977), "Estimating nonresponse bias in mail surveys", *Journal of Marketing Research*, Vol. 14, No. 3, pp. 396–402.

Ashton, K. (2009), "That 'internet of things' thing", *RFID Journal*, Vol. 22, pp. 97–114.

Baheti, R., and Gill, H. (2011), "Cyber-physical systems", *The Impact of Control Technology*, Vol. 12, pp. 161–166.

Bhamu, J., and Singh Sangwan, K. (2014), "Lean manufacturing: literature review and research issues", *International Journal of Operations & Production Management*, Vol. 34, No. 7, pp. 876–940.

Bibby, L., and Dehe, B. (2018), "Defining and assessing industry 4.0 maturity levels – case of the defence sector", *Production Planning & Control*, Vol. 29, No. 12, pp. 1030–1043.

Blanchard, D. (2010), *Supply Chain Management Best Practices*, 2nd ed., John & Wiley Sons, London, UK.

Bozarth, C., Warsing, D., Flynn, B., and Flynn, E. (2009), "The impact of supply chain complexity on manufacturing plant performance", *Journal of Operations Management*, Vol. 27, No. 1, pp. 78–93.

Brazilian National Confederation of Industry. (2016), "Industry 4.0: a new challenge for Brazilian industry", *CNI Indicators*, Vol. 17, No. 2. Available at http://elais.inf.utfsm.cl/wp-content/uploads/2017/08/t-ir-travassos-special_survey_industry4.0.pdf.

Bruce, M., Daly, L., and Towers, N. (2004), "Lean or agile: a solution for supply chain management in the textiles and clothing industry?", *International Journal of Operations & Production Management*, Vol. 24, No. 2, pp. 151–170.

Buer, S., Strandhagen, J., and Chan, F. (2018), "The link between Industry 4.0 and lean manufacturing: mapping current research and establishing a research agenda", *International Journal of Production Research*, Vol. 56, No. 8, pp. 2924–2940.

Cagliano, R., Caniato, F., and Spina, G. (2006), "The linkage between supply chain integration and manufacturing improvement programmes", *International Journal of Operations & Production Management*, Vol. 26, No. 3, pp. 282–299.

Cox, A., Chicksand, D., and Palmer, M. (2007), "Stairways to heaven or treadmills to oblivion? Creating sustainable strategies in red meat supply chains", *British Food Journal*, Vol. 109, No. 9, pp. 689–720.

Dalenogare, L., Benitez, G., Ayala, N., and Frank, A. (2018), "The expected contribution of Industry 4.0 technologies for industrial performance", *International Journal of Production Economics*, Vol. 204, pp. 383–394.

Dora, M., Kumar, M., and Gellynck, X. (2016), "Determinants and barriers to lean implementation in food-processing SMEs – a multiple case analysis", *Production Planning & Control*, Vol. 27, No. 1, pp. 1–23.

Dworschak, B., and Zaiser, H. (2014), "Competences for cyber-physical systems in manufacturing – first findings and scenarios", *Procedia CIRP*, Vol. 25, pp. 3–8.

Fang, E. (2008), "Customer participation and the trade-off between new product innovativeness and speed to market", *Journal of Marketing*, Vol. 72, No. 4, pp. 90–104.

Fatorachian, H., and Kazemi, H. (2018), "A critical investigation of Industry 4.0 in manufacturing: theoretical operationalisation framework", *Production Planning & Control*, Vol. 29, No. 8, pp. 633–644.

Flynn, B., Huo, B., and Zhao, X. (2010), "The impact of supply chain integration on performance: a contingency and configuration approach", *Journal of Operations Management*, Vol. 28, No. 1, pp. 58–71.

Forbes India. (2016), *Industry 4.0: the next industrial revolution demystified*. Available at http://www.forbesindia.com/article/nasscom-india-leadership-forum-2016/industry-4.0-the-next-industrial-revolution-demystified/42239/1. Accessed on March 6, 2017.

Frank, A., Dalenogare, L., and Ayala, N. (2019), "Industry 4.0 technologies: implementation patterns in manufacturing companies", *International Journal of Production Economics*, Vol. 210, pp. 15–26.

Frazzon, E., Albrecht, A., Hurtado, P., de Souza Silva, L., and Pannek, J. (2015), "Hybrid modelling approach for the scheduling and control of integrated production and logistic processes along export supply chains", *IFAC-PapersOnLine*, Vol. 48, No. 3, pp. 1521–1526.

Frohlich, M., and Westbrook, R. (2001), "Arcs of integration: an international study of supply chain strategies", *Journal of Operations Management*, Vol. 19, No. 2, pp. 185–200.

Ganzarain, J., and Errasti, N. (2016), "Three stage maturity model in SME's towards Industry 4.0", *Journal of Industrial Engineering and Management*, Vol. 9, No. 5, pp. 1119–1128.

Gjeldum, N., Mladineo, M., and Veza, I. (2016), "Transfer of model of innovative smart factory to Croatian economy using lean learning factory", *Procedia CIRP*, Vol. 54, pp. 158–163.

Goldsby, T., Knemeyer, A., Miller, J., and Wallenburg, C. (2013), "Measurement and moderation: finding the boundary conditions in logistics and supply chain research", *Journal of Business Logistics*, Vol. 34, No. 2, pp. 109–116.

Hair, J., Tatham, R., Anderson, R., and Black, W. (2014), *Multivariate data analysis*, Pearson Prentice Hall, Upper Saddle River, NJ.

Hermann, M., Pentek, T., and Otto, B. (2016), "Design principles for industrie 4.0 scenarios", *Proceedings of System Sciences (HICSS), 2016 49th Hawaii International Conference on IEEE*, Piscataway, NJ, pp. 3928–3937.

Hines, P., Holweg, M., and Rich, N. (2004), "Learning to evolve: a literature review of contemporary lean thinking", *International Journal of Operations and Production Management*, Vol. 24, No. 10, pp. 994–1011.

IBGE – Instituto Brasileiro de Geografia e Estatística. (2012), *Industrial Research in Brazil*, Available at https://ww2.ibge.gov.br/home/estatistica/economia/industria/pia/empresas/2012/defaulttabpdf.shtm. Accessed on May 7, 2018.

Jasti, N., and Kodali, R. (2015), "Lean production: literature review and trends", *International Journal of Production Research*, Vol. 53, No. 3, pp. 867–885.

Jasti, N., and Kodali, R. (2016), "An empirical study for implementation of lean principles in Indian manufacturing industry", *Benchmarking: An International Journal*, Vol. 23, No. 1, pp. 183–207.

Kagermann, H., Helbig, J., Hellinger, A., and Wahlster, W. (2013), "Recommendations for implementing the strategic initiative INDUSTRIE 4.0: securing the future of German manufacturing industry", Final report of the *Industrie 4.0 Working Group*, Forschungsunion.

Kampker, A., Deutskens, C., Heimes, H., Ordung, M., and Haunreiter, A. (2016), "Using e-mobility as an enabler for a fast and lean product development to optimize the return of engineering with the example of Lithium-ion battery", *Procedia CIRP*, Vol. 50, pp. 166–172.

Karnouskos, S. et al. (2014), "The IMC-AESOP architecture for cloud-based industrial cyber-physical systems", In: Colombo, A. et al. (eds.), *Industrial Cloud-Based Cyber-Physical Systems*, Springer, Cham.

Knoppen, D., Johnston, D., and Sáenz, M. J. (2015), "Supply chain relationships as a context for learning leading to innovation", *The International Journal of Logistics Management*, Vol. 26, No. 3, pp. 543–567.

Kolberg, D., Knobloch, J., and Zühlke, D. (2016), "Towards a lean automation interface for workstations", *International Journal of Production Research*, Vol. 55, No. 10, pp. 2845–2856.

Koufteros, X., Vonderembse, M., and Jayaram, J. (2005), "Internal and external integration for product development: the contingency effects of uncertainty, equivocality, and platform strategy", *Decision Sciences*, Vol. 36, No. 1, pp. 97–133.

Krajewski, L., Malhotra, M., and Ritzman, L. (2015), *Operations Management: Processes and Supply Chains*, Prentice Hall, New York, EEUU.

Kull, T., Yan, T., Liu, Z., and Wacker, J. (2014), "The moderation of lean manufacturing effectiveness by dimensions of national culture: testing practice-culture congruence hypotheses", *International Journal of Production Economics*, Vol. 153, pp. 1–12.

Landscheidt, S., and Kans, M. (2016), "Automation practices in Wood product industries: lessons learned, current practices and future perspectives", *Proceedings of the 7th Swedish Production Symposium SPS*, Lund, Sweden, October 25–27.

Lasi, H., Fettke, P., Kemper, H., Feld, T., and Hoffmann, M. (2014), "Industry 4.0", *Business & Information Systems Engineering*, Vol. 6, pp. 239–242.

Lee, J., Bagheri, B., and Kao, H. (2015), "A cyber-physical systems architecture for Industry 4.0-based manufacturing systems", *Manufacturing Letters*, Vol. 3, pp. 18–23.

Lee, J., Kao, H., and Yang, S. (2014), "Service innovation and smart analytics for industry 4.0 and Big Data environment", *Procedia CIRP*, Vol. 16, pp. 3–8.

Lewis, M. (2006), "Lean production and sustainable competitive advantage", In: Rhodes, E., Warren, J., and Carter, R. (eds.), *Supply Chain and Total Product Systems: A Reader*, pp. 306–326, The Open University, Milton Keynes.

Li, J., Tao, F., Cheng, Y., and Zhao, L. (2015), "Big Data in product lifecycle management", *The International Journal of Advanced Manufacturing Technology*, Vol. 81, No. 1–4, pp. 667–684.

MacDuffie, J. (1995), "Human resource bundles and manufacturing performance: organizational logic and flexible production systems in the world auto industry", *Industrial and Labor Relations Review*, Vol. 48, No. 2, pp. 197–221.

Malhotra, N., Birks, D., and Wills, P. (2006), *Marketing Research: An Applied Approach*, Pearson Education, London, UK.

Marodin, G., Frank, A., Tortorella, G., and Fetterman, D. (2019), "Lean production and operational performance in the Brazilian automotive supply chain", *Total Quality Management & Business Excellence*, Vol. 30, No. 3–4, pp. 370–385.

Marodin, G., Frank, A., Tortorella, G., and Saurin, T. (2016), "Contextual factors and lean production implementation in the Brazilian automotive supply chain", *Supply Chain Management: An International Journal*, Vol. 21, No. 4, pp. 417–432.

Marodin, G., Tortorella, G., Frank, A., and Godinho Filho, M. (2017), "The moderating effect of lean supply chain management on the impact of lean shop floor practices on quality and inventory", *Supply Chain Management: An International Journal*, Vol. 22, No. 6, pp. 473–485.

McCormack, K., Bronzo Ladeira, M., Oliveira, M. (2008), "Supply chain maturity and performance in Brazil", *Supply Chain Management: An International Journal*, Vol. 13, No. 4, pp. 272–282.

Mexican Ministry of Economy. (2016), *Crafting the Future: A Roadmap for Industry 4.0 in Mexico*, 1st ed., Government of Mexico, Mexico City. Available at www.gob.mx/se.

Meyers, L., Gamst, G., and Guarino, A. (2006), *Applied Multivariate Research*, Sage Publications, Thousand Oaks, CA.

Morris, D., Donnelly, T., Donnelly, T. (2004), "Supplier parks in the automotive industry", *Supply Chain Management: An International Journal*, Vol. 9, No. 2, pp. 129–133.

Moyano-Fuentes, J., and Sacristán-Díaz, M. (2012), "Learning on lean: a review of thinking and research", *International Journal of Operations & Production Management*, Vol. 32, No. 5, pp. 551–582.

Moyano-Fuentes, J., Bruque-Cámara, S., and Maqueira-Marín, J. (2019), "Development and validation of a lean supply chain management measurement instrument", *Production Planning & Control*, Vol. 30, No. 1, pp. 20–32.

Naim, M., and Gosling, J. (2011), "On leanness, agility and leagile supply chains", *International Journal of Production Economics*, Vol. 131, pp. 342–354.

Osterman, P. (1994), "How common is workplace transformation and who adopts it?", *Industrial and Labor Relations Review*, Vol. 47, No. 2, pp. 173–188.

Pérez, C., Castro, R., Simons, D., and Gimenez, G. (2010), "Development of lean supply chains: a case study of the Catalan pork sector", *Supply Chain Management: An International Journal*, Vol. 15, No. 1, pp. 55–68.

Podsakoff, P., and Organ, D. (1986), "Self-reports in organizational research: problems and prospects", *Journal of Management*, Vol. 12, No. 4, pp. 531–544.

Podsakoff, P., MacKenzie, S., Lee, J., and Podsakoff, N. (2003), "Common method biases in behavioral research: a critical review of the literature and recommended remedies", *Journal of Applied Psychology*, Vol. 88, No. 5, pp. 879.

Porter, M., and Heppelmann, J. (2015), "How smart, connected products are transforming companies", *Harvard Business Review*, Vol. 93, No. 10, pp. 96–114.

Power, D. (2005), "Supply chain management integration and implementation: a literature review", *Supply Chain Management: An International Journal*, Vol. 10, No. 4, pp. 252–263.

Qin, J., Liu, Y., and Grosvenor, R. (2016), "A categorical framework of manufacturing for Industry 4.0 and beyond", *Procedia CIRP*, Vol. 52, pp. 173–178.

Queiroz, M.M., and Telles, R. (2018), "Big Data analytics in supply chain and logistics: an empirical approach", *The International Journal of Logistics Management*, Vol. 29, No. 2, pp. 767–783.

Rauch, E., Dallasega, P., and Matt, D. (2016), "The way from lean product development (LPD) to smart product development (SPD)", *Procedia CIRP*, Vol. 50, pp. 26–31.

Ray, G., Muhanna, W., and Barney, J. (2005), "Information technology and the performance of the customer service process: a resource-based analysis", *MIS Quarterly*, Vol. 29, pp. 625–652.

Riet, C., Clerck, D., and Demeulemeester, E. (2015), *Lean Beyond Company Borders: Costs or Benefits?*, Faculty of Economics and Business of KU Leuven, Leuven, Belgium.

Rüßmann, M., Lorenz, M., Gerbert, P., Waldner, M., Justus, J., Engel, P., and Harnisch, M. (2015), "Industry 4.0: The future of productivity and growth in manufacturing industries", *Boston Consulting Group*, 9. Available at https://www.zvw.de/media.media.72e472fb-1698-4a15-8858-344351c8902f.original.pdf.

Sanders, A., Elangeswaran, C., and Wulfsberg, J. (2016), "Industry 4.0 implies lean manufacturing: research activities in industry 4.0 function as enablers for lean manufacturing", *Journal of Industrial Engineering and Management*, Vol. 9, No. 3, pp. 811–833.

Sanders, A., Subramanian, K., Redlich, T., and Wulfsberg, J. (2017), "Industry 4.0 and lean management–synergy or contradiction?", In *IFIP International Conference on Advances in Production Management Systems* (pp. 341–349), Springer, Cham.

Schmidt, R., Möhring, M., Härting, R., Reichstein, C., Neumaier, P., and Jozinović, P. (2015), "Industry 4.0-potentials for creating smart products: empirical research results", *Proceedings of International Conference on Business Information Systems* (pp. 16–27), Springer, Cham.

Schumacher, A., Erol, S., and Sihn, W. (2016), "A maturity model for assessing Industry 4.0 readiness and maturity of manufacturing enterprises", *Procedia CIRP*, Vol. 52, pp. 161–166.

Shah, R., and Ward, P. (2003), "Lean manufacturing: context, practice bundles, and performance", *Journal of Operations Management*, Vol. 21, No. 2, pp. 129–149.

Shah, R., and Ward, P. (2007), "Defining and developing measures of lean production", *Journal of Operations Management*, Vol. 25, No. 4, pp. 785–805.

Shariatzadeh, N., Lundholm, T., Lindberg, L., and Sivard, G. (2016), "Integration of digital factory with smart factory based on Internet of Things", *Procedia CIRP*, Vol. 50, pp. 512–517.

Shrouf, F., Ordieres-Meré, J., and Miragliotta, G. (2014), "Smart factories in Industry 4.0: a review of the concept and of energy management approached in production based on the Internet of Things paradigm", *Proceedings of Industrial Engineering and Engineering Management (IEEM)*, pp. 697–701. DOI: 10.1109/IEEM.2014.7058728.

Soni, G., and Kodali, R. (2013), "A critical review of supply chain management frameworks: proposed framework", *Benchmarking: An International Journal*, Vol. 20, No. 2, pp. 263–298.

Sousa, R., and Voss, C. (2008), "Contingency research in operations management practices", *Journal of Operations Management*, Vol. 26, No. 6, pp. 697–713.

Stentoft, A., and Vagn, F. (2013), "Evidence of lean: a review of international peer-reviewed journal articles", *European Business Review*, Vol. 25, No. 2, pp. 174–205.

Tabachnick, B., and Fidell, L. (2007), *Using Multivariate Statistics*, 5th ed., Allyn & Bacon, Needham Height, MA.

Taylor, D. (2006), "Strategic considerations in the development of lean agri-food supply chains: a case study of the UK pork sector", *Supply Chain Management: An International Journal*, Vol. 11, No. 3, pp. 271–280.

Tortorella, G., and Fettermann, D. (2018), "Implementation of Industry 4.0 and lean production in Brazilian manufacturing companies", *International Journal of Production Research*, Vol. 58, No. 8, pp. 2975–2987.

Tortorella, G., Marodin, G., Fettermann, D., and Fogliatto, F. (2016), "Relationships between lean product development enablers and problems", *International Journal of Production Research*, Vol. 54, No. 10, pp. 2837–2855.

Tortorella, G., Marodin, G., Miorando, R., and Seidel, A. (2015), "The impact of contextual variables on learning organization in firms that are implementing lean: a study in Southern Brazil", *The International Journal of Advanced Manufacturing Technology*, Vol. 78, No. 9–12, pp. 1879–1892.

Tortorella, G., Miorando, R., and Marodin, G. (2017a), "Lean supply chain management: Empirical research on practices, contexts and performance", *International Journal of Production Economics*, Vol. 193, pp. 98–112.

Tortorella, G., Miorando, R., and Tlapa, D. (2017b), "Implementation of lean supply chain: an empirical research on the effect of context", *The TQM Journal*, Vol. 29, No. 4, pp. 610–623.

Tortorella, G., Miorando, R., Caiado, R., Nascimento, D., and Portioli Staudacher, A. (2018a), "The mediating effect of employees' involvement on the relationship between Industry 4.0 and operational performance improvement", *Total Quality Management & Business Excellence*. Available at https://www.tandfonline.com/doi/abs/10.1080/14783363.2018.1532789.

Tortorella, G., Miorando, R., Fries, C., and Vergara, A. (2018b), "On the relationship between Lean Supply Chain Management and performance improvement by adopting Industry 4.0 technologies", *Proceedings of the International Conference on Industrial Engineering and Operations Management* Paris, France, July 26–27.

Vitasek, K., Manrodt, K., and Abbott, J. (2005), "What makes a lean supply chain", *Supply Chain Management Review*, Vol. 9, No. 7, pp. 39–45.

Weyer, S., Schmitt, M., Ohmer, M., and Gorecky, D. (2015), "Towards Industry 4.0-Standardization as the crucial challenge for highly modular, multi-vendor production systems", *IFAC-Papersonline*, Vol. 48, No. 3, pp. 579–584.

Wollschlaeger, M., Sauter, T., and Jasperneite, J. (2017), "The future of industrial communication: automation networks in the era of the internet of things and industry 4.0", *IEEE Industrial Electronics Magazine*, Vol. 11, No. 1, pp. 17–27.

Yusuf, Y., Gunasekaran, A., Adeleye, E., and Sivayoganathan, K. (2004), "Agile supply chain capabilities: determinants of competitive objectives", *European Journal of Operational Research*, Vol. 159, pp. 379–392.

Zacharia, Z., Plasch, M., Mohan, U., and Gerschberger, M. (2019), "The emerging role of coopetition within inter-firm relationships", *The International Journal of Logistics Management* (forthcoming).

Zhao, X., Huo, B., Selen, W., and Yeung, J. (2011), "The impact of internal integration and relationship commitment on external integration", *Journal of Operations Management*, Vol. 29, No. 1–2, pp. 17–32.

chapter six

Fundamental changes in the organisational processes

Industry 4.0 case study

Vasja Roblek
Faculty of Organisation Studies, Novo Mesto, Slovenia

Ivan Erenda
TPV Group, Novo Mesto, Slovenia

Maja Mesko
University of Maribor and University of Primorska, Slovenia

Contents

6.1 Introduction

This chapter is focused on crucial changes during the fourth industrial revolution that affects the changes in organisational processes. Reorganisation based on technological advances has profoundly changed organisational processes, organisational forms, organisational relationships and work processes throughout industrial history. The impact of technological development on the organisational changes across the three industrial revolutions is presented at the beginning. Particular importance is given to the fourth industrial revolution or Industry 4.0, which has made significant changes to products and production system through processes, design, operations, and services (Ślusarczyk, 2018). The smart factory, as one of the most relevant results of Industry 4.0, is presented in the framework of the cybernetics theory.

The main aim of this chapter is to study fundamental changes in the transformation of Slovenian smart factory TPV Ltd. Thus, the firm has transformed the once-classic production of an auto parts factory into a smart development centre for business management, logistics and manufacturing. The core business has been transformed from automation to robotics as well.

In our chapter, a systematic approach to technological changes in manufacturing is presented at the beginning. After that, the consequences of informatisation, which will have an impact on changes in an internal and external company environment, were anticipated. The discussion is finally presented by highlighting the theoretical and practical implications of the research.

6.2 *Industrial revolutions and their impact on organisational changes and adoption*

The economic historians defined three industrial revolutions in the modern period of society and capitalism, and we have just begun entering the fourth industrial revolution. Table 6.1 shows the industrial revolutions and its main features.

The first industrial revolution started in 1780 is based on the innovation of the steam engine, which enabled the transition from the manual work to the era of machines. The second industrial revolution (around 1860) triggered changes in the technological matrix in transport (railroad, steamer, and later internal combustion engine), in new industries (chemical, electrical industry) as well as the improvement of old industries (steel production).

In the second half of the 20th century, the third industrial revolution occurred (began in 1960). The invention of the transistor and microprocessor had an essential influence on raising the telecommunication and computing technology (the process of computerisation). Hence, new process control was not about physically restricting options – as in technical control – neither about bureaucratic top-down order. Its primary goal, instead, was the development of computer monitoring combined with human control in the production process. Automation processes in

Table 6.1 Industrial revolutions attributes

Changes through industrial revolutions	Ist and IInd industrial revolution	IIIrd industrial revolution	IVth industrial revolution
	1780 ↔ 1860 ↔ 1960	1960 ↔ 2011	2011 ↔ 2040
Organisational type	Taylor's hierarchical system; Weber's bureaucratic model	Flat organisation, networking	Lean and virtual organisation, networking
The importance of capital and its transformation	Financial capital, production facilities, material goods	Financial and social capital in the form of talented people who do create jobs, and there are extendible to greater independence from the culture of the older generations	Social capital, with an emphasis on sustainable development of natural resources, aspirations for social justice and less stressful working environment
Sector changes	Industrial sector	Sectoral dispersion, clusters of interrelated activities	Social networks, clustering of different activities with each other in order to find new solutions, outside of the frame work of each science
The form of human processes	Productivity	Profitability	Innovativeness
Technological processes	Automatisation, computerisation	Digitise/ digitalisation	Informatisation and digitise
Core activities for the growth and operation of all types of organisations	Quantity	Quality	Creativity and innovativeness
Energy resources	Economy dependent on fossil fuels	Development model still based on fossil fuels but it was occurred the sustainable resource development model	Renewable energy production (solar energy, biomass, thermal power, hydrogen cells)

(*Continued*)

Table 6.1 (Continued) Industrial revolutions attributes

Changes through industrial revolutions	Ist and IInd industrial revolution	IIIrd industrial revolution	IVth industrial revolution
Strategic planning of work processes	Long-term business planning, little change and this only at the operational levels well as the lay out of production in one place	Short-term business planning, more rapid changes, both at operational and strategic levels of operations	Planning operations due to constant changes, which requires permanent change at all levels of business, is not possible
The types of works success	The success achieved on the basis of related competitive advantages set of resources and skills; the workforce has been trained for each function in the production process	Success depends on the ability to learn and adapt organisations and personnel to market conditions	Success depends on the ability to develop and use intuition
Competitiveness	Competitiveness is achieved at the expense of lowering operating costs, lower taxes and lower labour costs	Competitiveness is achieved by moving companies in places where talented personnel, adequate business environment, which allows developed business and social infrastructure, which facilitates the development of business partnerships and providing trained personnel with ideas	Developed infrastructure environment with well-developed public transport, fast and affordable web links and cultural environment that allows personality development
Knowledge management	Knowledge concentrated in management	Knowledge is concentrated in the organisation's employees	Knowledge stored outside organisations free access, open innovations

Table 6.1 (Continued) Industrial revolutions attributes

Changes through industrial revolutions	Ist and IInd industrial revolution	IIIrd industrial revolution	IVth industrial revolution
Labour market	The labour market has been largely regional	The labour market has become global, present many physical relocation of staff worldwide	Global labour market, virtualisation business to learn and work from home
Government, public apparatus and non-profit sector	Economic development led government; large state apparatus meant good public service; Webers' bureaucracy	Post-bureaucratic model; automatisation and computarisation of public administration; with the internet was launched the –e-Government and –e-Democracy; Partners networking were established between government apparatus and non-profit sector	Reduction of the state apparatus to a minimum, taking over the functions of government by the non-profit sector with the support of private organisations and the transition to e-law, e-education, e- health, smart governance, etc.

Source: Modified from Roblek et al. (2014).

organisations had encouraged researchers in 1960s of the 20th century to begin with the researching of the technological and automatisation impact on the workforce and economic progress. The essential questions were in this time how automatisation effects on employment levels and organisational efficiency. At the end of 1960, we faced with the beginning of automatic processing of a large number of data using large, powerful computers like the IBM 370. They allowed simultaneous work of several users, who were connected to them through the terminals and later with the microcomputers. During this period, the use of computers was limited to the operational level of the administration's operations and administrative, technical and simplified professional tasks, such as the production of structured documents, business accounts and the aggregation of financial and operational results (Coe and Borthick, 1984; Kelly, 1978).

It was a mass processing of documents, applications supported separate procedures, which caused the problem of redundancy of data (Bekkers and Frissen, 1992). These systems were extremely inflexible because most applications were such that they had to be rewritten entirely in the event of a new requirement (Kelly, 1978). There were no organisational changes in the administration during the automation period. The administrative procedures remained virtually unchanged, only individual routine works, which were previously performed manually, were automated. The purpose of using technology during this period is to reduce the actual costs of paper operations and to use the information as a by-product (Bellamy and Taylor, 1998). The organisation of the companies and public administration itself remains bureaucratic. The process of manufacturing depends at the beginning of the computerisation on a human operator. The role of the human operator was to monitor and respond to the production process based on the information produced by computers in an open-loop form of feedback that would eventually become closed-loop automation (Schaupp and Diab, 2020).

The third revolution resulted in change from analogue and electronic technology to digital technology that emerged in 1980 and continues to develop today. The invention and production of miniaturised material led to the increasing researches in space technologies and biotechnology. Two crucial inventions of the third industrial revolution are also automatons—programmable logic controllers (PLCs)—and robots that enable the high-level automation in production. The main feature of this period is mass production and extensive use of digital logic circuits, and its derivative technologies, including a computer, a mobile phone, fax machine and internet (Schwab, 2016, pp. 6–8; Taalbi, 2018, pp. 2–3).

The process of computerisation and implementation of the telecommunication technologies throughout the white-collar workforce began at the end of 1960 and continued in 70 and 80 years of the 20th century. The process influenced the social scientists with the consideration of how these information technologies have effects on the social experience of individual users (e.g. the quality of their working environment, job satisfaction). These researches were given a contradictory finding (Attewell and Rule, 1985, p. 1185). One of the socio-economical theoretical conclusions is that information technology can increase productivity, but influenced the user's degradation of the work-life. More qualified employees will replace less educated, unskilled and unexperienced employees. Information technology was at that time able to handle this in two ways. First, it was able to integrate direct specific skills and knowledge, considering that the user does not need to have as much of this knowledge and skills (Allen and Hirshon, 1998). Second, it was able to complement Taylor's hierarchal system with the fragmentation of an integrated job into relatively autonomous components. The goal of using information technology was to incorporate users (employees) skills and to reintegrate information produced through fragmented jobs (Glenn and Feldberg, 1977; Gregory and Nussbaum, 1982; Murphree, 1984).

In contrast to the previously presented facts, some researchers like Giuliano (1982) and Turner (1984) have highlighted the fact that the use of computers allows greater sophistication than everyday work. Adler (1983) pointed out that routine tasks will become automated, which will result in the dismantling or cessation of jobs requiring a low level of skills. Poppel (1982) was preparing research in 15 U.S. organisations. The results showed that a salesperson was wasting his time on travel, missing contacts, finding out information and filling out forms. On the other side, managers and professionals spent their time on administration (clerk) work and meetings. Poppel concluded, according to the results of his analysis, that problem with time wasted could be partly rescued with the office automation technology. Attewell (1985) research showed, on concrete cases, how the process of computerisation eliminated some of the routine work of insurance auditors (e.g. calculating of deductibles or identifying of potential duplicate payments). Thus, computerisation allowed auditors that instead of everyday work could begin paying more in-depth with the making decisions about dubious claims. Zuboff (1988) claimed that computerisation is making activities more abstract, and this is resulting in adding intellectual content to work.

Zuboff (1985) discussed how information technology refers to computer – mediate work. It is a process where a task is accomplished through the medium of the information system instead of direct physical contact with the task. Zuboff (1988) described two types of existing information technology. The first type is automating technology, which is designed and implemented to lessen or deskill the processes which make up the work. The second type represents information technology which can be designed and implemented to improve the processes which define the work. The importance of information technology is in removing the most boring, repetitious, dangerous and mindless task. Such information technology allows employees to devote their time to creative, challenging, intellectual and satisfying aspects of their work. The studies from 70 and 80 years of the 20th century showed that computerisation was influenced by the changing nature of work (work schedules, staffing levels, location and structure of departments), organisational structure (reduction of the hierarchy levels), decentralisation of work and responsibilities, people's morale and organisational culture (e.g. relations with co-workers and managers). All of this adds up to the notion that technology, per se, is a primary factor driving organisational behaviour (Braverman, 1974; Davis, Bagozzi, and Warshaw, 1989; Hirschhorn, 1984; Shaiken, 1988).

In the 1980s and early 1990s, affordable computers and information equipment contributed to further organisation modernisation. New generations of programming languages (e.g. C+, SQL), relational databases and LANs have enabled the integration of previously separate tasks, the interdependence of organisations from information, greater transparency of business and decision-making. In the period of computerisation and digitalisation, the trend from individual computer workplaces was replaced by multipurpose workplaces with access to programs from local and central units (Ceruzzi, 2003).

Computerisation means using a computer to do something that was before done by people or machines (Cambridge Dictionary, s.v.). For example, information systems influenced the computerising the medical records (e.g. Chow et al., 2012), computerised prescriptions are used instead of a hand-written prescription (Guchelaar and Kalmeijer, 2003).

Word digitalisation cannot be used as a synonym for the computerisation (Paulin, 2019, p. 150). In business administration digitalisation often means adoption and increased use of the information and communications technologies (Katsikas and Gritzalis, 2017). Other meanings of the word digitalisation are connected with the process of converting something to digital form (e.g. digitalisation of payments, digitalisation of the printed documents and books and the transformation of the analogue signals in a digital representation) (Merriam-Webster, 2019). According to the Paulin (2019, p. 150): "computerisation implies closer proximity to software for directly controlling hardware (e.g. company information system controls their hardware)".

The word informatisation is using for the processes of the introduction and adoption of information systems, technologies, for instance, computerisation of the company, public apparatus and business.

Described three words (informatisation, digitalisation and computerisation) are often used for the same meaning. In this text, the terms informatising/informatisation/informatised is used only in the context of fourth-generation systems (systems developed and launched in the fourth industrial revolution). The results of the informatisation are informatised entities, which can be created, controlled and transformed within cyberspace (Paulin, 2019, p. 148).

The development of the internet can be classified among the most significant innovations in the second part of the 20th century. The internet and new digital technologies offer organisations better opportunities for acquiring strategic priority positions. Since the beginning of the 1990s, the web has a significant impact on the economic, business and sociopolitical aspects of our lives. In the business field, it has contributed to the development of e-commerce, and in public apparatus, it has introduced concepts such as e-Gov and e-Democracy. However, the critical question is how to use the web to increase economic value.

Internet technologies, e-Commerce and e-Gov, are joined by digital technology. Today, digital media is used in creative expression in digital art, science, technology and business, and enables expression, communication, virtualisation, social interaction and education. Their expansion was caused by the phenomenon of the knowledge economy (e.g. digital or internet economy), which has emergence together with the internet era.

Rapid technological developments, cheaper computer equipment, Web 2.0, social media and increased availability of broadband internet have changed people lifestyle and work style. Technical capabilities, new internet technologies have a significant impact on sociopolitical developments. Internet technologies had an essential role in the economic field, as they significantly influenced the creation of new jobs in the crisis between 2008 and 2010. Thus, for each of the two lost jobs, it created five new ones (Pélissié du Rausas et al., 2011).

In the knowledge economy, from 2008 onward, economic and social activities have been globally interconnected, enabling technology platforms, such as the internet, mobile and sensory systems. The knowledge economy has increased interaction between technological development and stimulated changes in formal and (in)formal social relations (Castells, 2010; Etzkowitz and Leydesdorff, 2000; Guile, 2010). Information and communication technology enables constant interactivity, connectivity, transparency and productivity skipping (Kaplan and Haenlein, 2010, p. 61). All mentioned changes lead to further changes in world markets, which are reflected in various ways, in the changed consumer behaviour and consequently in the development of new business models (Roblek et al., 2013, p. 558). All these factors have taken the lead in the field of ICT in developed and less developed economies. In the transition period from the new to the innovation economy, technology and economic growth have become inseparable (Kaplan and Mikes, 2012, p. 51). With the development and increasing use of digital technologies, the single European market is mutually reinforcing.

The rapid development of information and communication technology and the political desire to launch the European post-crisis economy from 2008 to 2010 prompted the German government to prepare a proposal Industry 4.0 for the development of a new concept of German economic policy, which is based on high-tech strategies and influence on changes in business, government, healthcare and consumption (Mosconi, 2015; Roblek, Erenda, and Meško, 2018). This political proposal was a foundation for the launching of the fourth industrial revolution (2011–), also known as Industry 4.0. The concepts of Industry 4.0 develop the cyber-physical systems (CPSs), the Internet of Things (IoT), the Internet of Services (IoS) and forthcoming 5G (Lasi et al., 2014; Ning and Liu, 2015). It is based on constant communication via the internet that allows a continuous interaction and exchange of information not only between humans (C2C) and human and machine (C2M), but also between the machines themselves (M2M) (Cooper and James, 2009). All must be connected – hardware, data and connections, and the common denominator is software. Nowadays, organisations and governments are building software environments, the so-called platforms that will run all the processes of the human lives and on which the majority of the business will be based. Digital data is open-sourced and available free. The smart technologies (e.g. phones, tablets, watches), applications, sensors and networks have generated the data in real-time from human professional (e.g. manufacturing cloud) and private environment (e.g. smart house) (Roblek, Meško, and Krapež, 2016).

The fourth industrial revolution changed the meaning of the phrase network from a definition of connection to a source for innovation (Hanna, 2016). The data become a substantial value for business and society. Data and data management become core components for success and competitive advantage in many industrial sectors, value chains or organisational processes and thus a decisive factor for production, alongside labour and capital. This is the reason why organisations need to connect or integrate existing data infrastructures, which will increase the quality of existing data and decrease communication and interoperability costs.

Zuboff (2019) reminds us of "surveillance capitalism", in which one-sided analysis made based on free and raw material is captured (e.g., workplace cameras, street cameras, web users' views). Although some of this information is used to improve services, others have been declared as a proprietary behavioural surplus that feeds into advanced manufacturing processes. Last but not least, they are traded on data markets, which Zuboff calls behavioural futures markets.

The transformation processes are seemingly independent, and the analysis of the real state clearly shows that they are intimately connected. For every phase of transformation, the specific social rules are applied (e.g. legal, economic and ethical).

The new evolutionary phase is linked to the development of the social superstructure and occurs when appropriate conditions are created for it in the broader social environment, like the level of information knowledge development, awareness of the individual and attitude towards the environment. With the emergence of a new evolutionary phase, the characteristics of the former are blended with its characteristics. Perhaps it is time for us to stop watching social systems as something artificial, but accepting them as one of the natural forms of coexistence and development. That will give us a much more comprehensive range of possibilities in the field of socio-economic models. The focus of management and leadership spontaneously moves from functions and processes closer to the person to direct relationships.

In the fourth industrial revolution, the source of creating new value becomes creativity, and the leverage for creating the winning factors in the market is innovation and intuition (Kuula, Putkiranta, and Toivanen, 2012, p. 117). Development of knowledge, professional attitude towards work and the acquisition of new information and communication technologies become a stable basis of creativity and innovative processes as a way to increase efficiency and effectiveness (Bisson, Stephenson, and Vigurie, 2010; Kaplan and Mikes, 2012, p. 52).

Today, innovative environments are a challenge. Besides, leaders are increasingly concerned about the proper working environment and the rules of the game, since only efficient use of intellectual resources allows continuous improvements. For this reason, the behaviour of business and governmental systems is changing at all levels of leadership, management and implementation. Organisations need creative employees who are involved in the processes of strategic thinking and who are capable of new values, creativity and innovation to pass on to the remaining colleagues. Thus, human beings become the most crucial source that needs to be managed effectively.

The characteristic of the present time is innovation-based entrepreneurship and governmental sector based on development and research, deregulation, increased risk capital financing and international protection of intellectual property (Braunerhjelm and Henrekson, 2016). It is vital to network business as a strategic tool that enables the acquisition of knowledge and information and links people with knowledge in a modern knowledge society. The corporate network provides synergies in the co-management of information, knowledge and human resources. Knowledge and information are a crucial factor in success in the third and fourth industrial revolution (Donate and de Pablo, 2015).

From the point of optimisation of resources, the organisation must make as much as possible, as this means lowering the costs per unit produced and, thus, greater efficiency. Higher productivity with cost optimisation means a competitive edge.

From the position of value and value system, it is essential to understand the contemporary direction: striving for a balance between business and personal life, a creative environment and the possibility of self-realisation. The new evolutionary phase is linked to the development of the social superstructure and only occurs when appropriate conditions are created for it in the broader social environment, these are level of information knowledge development, awareness of the individual, attitude towards the environment. With the emergence of a new evolutionary phase, the characteristics of the former are blended with its characteristics. Perhaps it is time for us to stop watching social systems as something artificial, but accepting them as one of the natural forms of coexistence and development, what gives us a much more extensive range of possibilities in the field of socio-economic models.

Development of knowledge, professional attitude towards work and the acquisition of new information and communication technologies become a stable basis of creativity and innovative processes as a way to increase efficiency and effectiveness (Bisson, Stephenson, and Vigurie, 2010; Kaplan and Mikes, 2012, p. 52).

6.3 Methods

6.3.1 Designing a single case study

An exploratory qualitative study was used as part of the study. As the term suggests, an exploratory study is concerned with exploring a phenomenon. In the specific case, the smart factory TPV represents a case study. The Group TPV is, in addition to its core business, focused on R&D technologies for smart factories. In their company, TPV Ltd. developed their own automated guided vehicle (a tunnel model) whose key advantage is its low height. Moreover, TPV offers its customers a comprehensive automatic supply to workstations. Automation of internal logistics with automatic guided vehicles (AGVs) is undoubtedly an excellent opportunity for companies to optimise their logistics processes (Lekše, Sluga, and Rajšelj, 2019a).

Thus, we used a single case study approach to collect, organise and analyse data received from seven employees. The case study allows the thematic analysis of cases based on information obtained in the field (Patton, 2002). A qualitative case study is based on the specific collection, organisation and analysis of data. Thus, it represents an analytical process. The purpose of the case study is to obtain comprehensive, systematic and in-depth information on the research phenomenon. The result of the analysis is a case study.

The study is designed by the general case study method (Easterby-Smith, Thorpe, and Jackson, 2012). Consequently, such changes within the company contain elements of a research and descriptive case study of Yin (2009). Consider that an appropriate combination of both grounded theory and case studies represent an appropriate research approach to the design of a study model that makes it possible to put research results into practice. It makes sense when we want to understand in-depth and explain a phenomenon or happening (Yin, 2009).

6.3.2 Data collection

Qualitative research does not set a rule for how many people should count the sample. The size of the sample thus depends on the purpose of the research, its needs, applicability, the importance of credibility, time available and resources (Patton, 2002). The aim of qualitative research is not to find findings with a representative sample that can be generalised to the basic set, but to get to the fullest possible extent the person, institution, phenomenon and group studied. Therefore, in the context of qualitative research, we focus on examining a small number of cases, or only on a case by case basis (one person, one institution, event and group). It is essential that qualitative research also provides findings that can be generalised to the general public. Thus, even in the case of a case study, it is possible to formulate a theory that can later be tested in other individual cases to determine the general principles.

The sample in the qualitative research should be relevant to the research. The researcher must be sure that he/she will obtain with the selected units the necessary information to answer the research questions (Schutt, 2012).

In the research, dedicated sampling was used. Thus, each unit was selected for a specific purpose.

6.4 Smart factory as a research paradigm

In the research concept of a smart factory system and its influence on organisational changes and adoption, it is crucial to explore a living system that has its own "will", regardless of its level of development (technology development, the degree of reorganisation of the organisation, leaders and management readiness to transition from the "old" manufacturing into smart manufacturing and cyberspace, self-organised system). Living systems such as a smart factory are organised organically, consequently, are more challenging to manage. Their interactions with their environment are complicated, and it is impossible to do more than one step ahead predictions. It is a system of increasing control capability, developed by second-order cybernetics; the cybernetics focused on the understanding of biological and social complexity as well as its control. According to this orientation, it is understandable that the other type of cybernetics is more focused on creating new forms (morphogenesis) and a positive feedback loop (Dominici and Roblek, 2016).

In the case of research phenomenon presented in this chapter, which is the smart factory, it is vital to understand the role of industrial IoT, cyber-physical production systems (CPPSs), and production systems consisting of one or more CPS. The CPS presents a physical object with an embedded system containing a control processing unit (computer power), the industrial cloud which can store, analyse and share data with form network connectivity (Mabkhot et al., 2018). In the field of smart manufacturing, the goal of R&D is to reach the self-organisation with automatic machine configuration and process optimisation which is enabled in the form of decentralised production control.

The innovative production software technology is changing the shop floor management by introducing the advanced systems including lean management philosophy. The enterprise resource planning (ERP) software on the planning level (top floor) is using objective performance data capture from all factory assets (shop floor) in real-time. Manufacturing execution system (MES) has to improve the production processes. It aims to enable the connection between production data with ERP data from corporate planning like provided resources, consumer requirements and expectations (Gruber, 2013; Oesterreich, 2016).

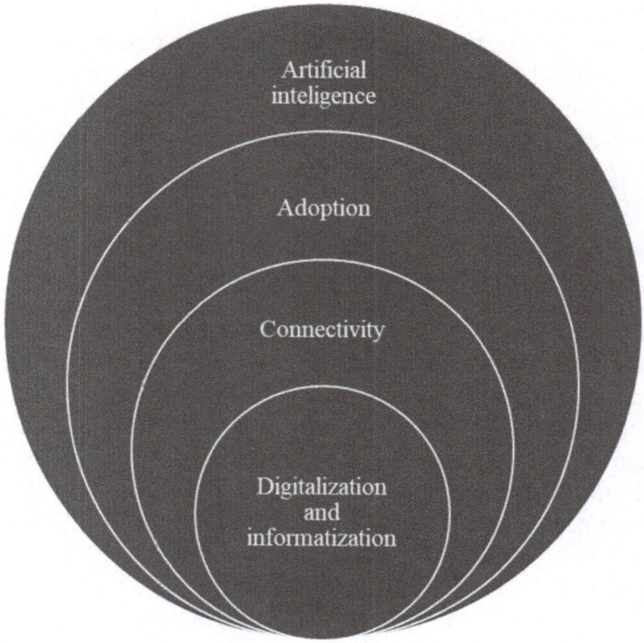

Figure 6.1 Smart factory technological development. (Adapted from Lekše, Sluga, and Rajšpelj, 2019a.)

6.4.1 *Development of the production system*

As can be seen from Figure 6.1, the starting point for technological development in the smart factory production systems is digital transformation, measurement, digitisation and informatisation of everything related to production systems. The basis for evaluation, integration and optimal process management are process-related data. Data are obtained from measurements performed by different sensors. Smart sensors with an integrated microprocessor play an important role in the field of measurement, allowing its immediate digitalisation. Integrated smart sensors provide the performance of logical functions, two-way communication, and adaptation to environmental changes, decision-making, self-calibration and self-testing at start-ups. The sensors are getting smaller and more user-friendly.

In TVP Ltd., the digitisation enables to connect smart factories with other elements of the smart infrastructure – people, machines and products. It is about linking up the entire value chain throughout the lifespan. People are involved in the roles of customers, constructors, technologists, managers and enhancers, repairers and analysts.

Connectivity enables them to adapt their systems to the needs of our customers in all aspects, specific requirements, quantities, deadlines and delivery points.

Their R&D on the smart factory is going in the way of an autonomous, self-organised system; it will make decisions on their own, correcting any malfunctions. These are complex systems based on artificial intelligence, stochastic models.

According to the opinion of their R&D engineers, it is the firm primary goal to implement the concept of smart factories. It will be necessary to overcome some of the challenges that will require thoughtful and coordinated action by crucial industry development players. The main challenges they are facing are standardisation, security and IT infrastructure. The real establishment of mentioned elements in the broader industrial environment will take several years, which is why some prefer to use the word evolution instead of the industrial revolution.

6.4.2 *Simplify complicity*

Complicity simplifying is one of the significant R&D goals of the group. Their opinion is that as development will go forward with high intensity, product development and product life cycles will be shorter. Solutions, products and machines offer users more and more functionality, automatic and autonomous operation. The equipment is increasingly smart and connected. The solutions are also more complex and often challenging to handle, even for advanced users. That is why we need to be especially aware that the following three user requirements are always crucial in concrete applications (Lekše, Sluga, and Rajšpelj, 2019a):

- Reliability of operation
- Efficiency (technical and economic)
- Simplicity

For solution users, the most important thing is that the systems work reliably and robustly, and that they are not sensitive to changes and different external influences. In making a decision, efficiency, technical and cost-effective and competitive requirements are always one of the deciding factors. The third important factor is the ease of operation and use, the importance of which will continue to increase in the future. With the intensive introduction of automation and smart systems into production, the circle of users of systems and solutions will increasingly expand to employees with lower levels of expertise. Therefore, the solutions must be easy to use and designed for intuitive management.

Due to the development of technologies, a wide range of solutions and possible applications, choosing the right, the optimal solution for a specific case is an extremely important and challenging.

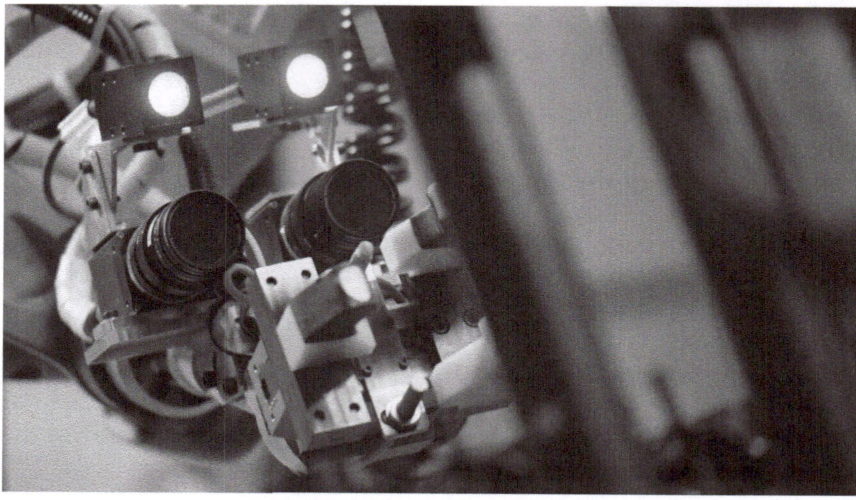

Figure 6.2 TPV Optimatic 160 PRO automatic guided vehicle for production logistics automation. (Adapted from TVP Ltd, with permission.)

In the case of automation of internal logistics with AGVs, it is critical to know where in the industry and why the solution is optimal and undoubtedly represents an excellent opportunity for companies to optimise their logistics processes and improve efficiency (Lekše, Sluga, and Rajšpelj, 2019a). Figure 6.2 illustrates the AGV.

TPV Ltd. develops and markets solutions in the field of automation of internal logistics in production and warehouses with AGV. It is a complete solution of automated, synchronous supply of jobs that do not require operators to perform. The smart solution is the central control system. It has several modules that handle multi-vehicle traffic, communication with other elements of automation and signalling, control and automatic recharging of batteries, maintenance and notification and advanced layout. Only material-driven vehicles can be manipulated, which can be in different versions: tunnel, tractor, roller track and forklift. An additional competitive advantage of our solutions is the innovative automatic guided Optimatik 160, which has been developed intending to offer the best solutions to the customer. It is a vehicle (tunnel type) whose key advantage is its low height. The vehicle is only 160-mm high, which is unique in the market. The height of the AGV is less than the height of the standard wheels used for industrial trolleys, so no trolley alterations are required from customers. As they said, "Our internal logistics automation solutions are smart factory solutions".

Automation of internal logistics with AGV has many advantages for users in various fields. Supply of manufacturing jobs is more efficient, synchronous and more reliable, which is reflected in increased productivity. The chances of human error are reduced. The need for human labour in logistics and thus labour costs are also reduced. The system provides 24-hour operation, 7 days a week. A significant advantage is the significantly lower power consumption compared to conventional forklifts – all these results a quick return on investment, which is usually less than 1 year. The flexible and modular design of the system makes it easy and quick to make changes and make various upgrades. In the field of occupational safety and health, the risk of injury to persons, equipment and material is significantly reduced. Both the number of incidents and the severity or consequences of injuries to the persons involved are reduced.

The software also allows additional software modifications. Therefore, the layout changes such as new roads, changing delivery and pickup points can be easily performed.

The whole system is operated by the central control system, which consists of modules for traffic management of several AGVs, communication with other elements of automation and signalling, control and automatic charging of batteries, error management and notification and advanced planning and scheduling.

The first module is the traffic management, which in the case of crossing the path by several AGVs determines the priority of moving or avoiding the stalemate of two or more AGVs. The stalemate means that two or more AGVs are encountered and stopped at a security scanner.

Another unique feature is intelligent control of trailer and cart presence in the pick-up zone while the sensor is not activated, but the workstation needs to be supplied. In this situation, the AGV automatically scans the line, detects the cart presence and transports it to its destination.

The second module is called supporting elements of automation, additional signalling (PLC). PLC enables the construction of additional automation according to the customer specification. It means that in conjunction with AGV, it can communicate and manage other internal logistics equipment, such as conveyor belts, lifting systems, cogwheel drives, intelligent storage units, self-service racks, various locking elements, external security signalisation, additional signalling on the route, visual notification, etc. PLC supports all relevant protocols (OPC, TCP/IP Modbus, etc.). The PLC interacts with the surroundings and the AGV system (sensors, confirmation keys, automatic battery charging, etc.).

In addition to two modules described above, the battery management module continually monitors the status of the battery and at any waiting time charges them. In this case, the standard charge with stopping AGV for a few hours (4–7 hours) is eliminated. This allows the AGV to operate autonomously.

The Plant Maintenance and Notification reports and records all AGV stops that occur due to unforeseen obstacles, mechanical or electrical errors. The system reports an error via the computer screen, SMS notification or application on the smartphone. Information includes the type of error and location for quick action.

Advanced planning and scheduling module allow us to automate material handling in production and automate supply of workstations according to the pull system.

One of the successful innovative solutions in the field of logistics automation with AGV is also the intelligent Canban system. The intelligent Canban system is a supply job-care system without operators. It works on the principle of "pull". Supply chute, that is, intelligent storage (I-storage), is equipped with sensors in the workplace, which recognise the minimum stock (minimum number of boxes), or the maximum stock (maximum number of boxes) on the supply chute. With a minimum stock, the I-mailbox sends a request via Wi-Fi communication to the central control system (CCS). The CCS accepts the request and sends a request to the AGV with a message indicating which collection (i.e. the synchronised number of pieces of material for the manufacture of finished products) delivers to the workplace or to the I-storage, which triggered the request. AGV autonomously transports a collection of material boxes from the picking zone to the I-storage, where it automatically unloads and completes the stock on the I-storage and takes empty boxes. AGV returns the empty boxes to the location thus planned, it is unloaded and returned to the collection zone, where it waits for a new request.

The CNS can inform the warehouse operator via a traffic light or LCD screen to complete the stock in the collection zone.

CNS can be linked to the following systems, which enable an even higher level of intelligent implementation of the I-KANBAN system (Lekše, Sluga, and Rakšelj, 2019b):

- System planning of MRP material
- Warehouse Management System (WMS)
- Manufacturing Execution System (MES)
- Advanced planning and scheduling

A higher level of communication enables them to automatically record the material (OK or NOK) at the workplace, regular inventory management of the incoming, on-going and outgoing stock, timely information on the replacement of the type of product and consequently ensure the minimum time for the replacement of the tool and the timely completion of the I-storage with new materials (Lekše, Sluga and Rakšelj, 2019b). The processes are shown in the following figures: Figure 6.3 presents the smart factory module for autonomous commissioning of shipping pallets; Figure 6.4 shows automatic control of the

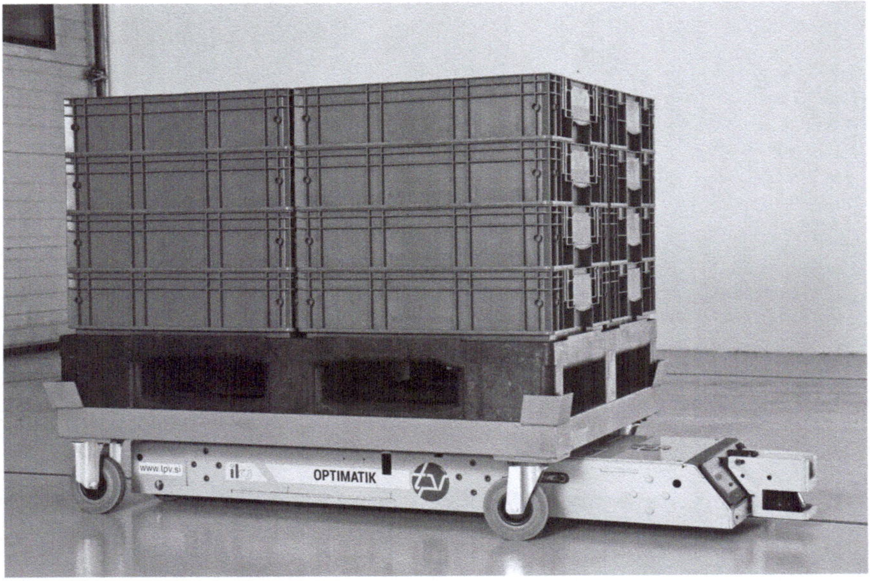

Figure 6.3 Smart factory module for autonomous picking of dispatch pallets. (Adapted from TVP Ltd, with permission.)

Figure 6.4 Automatic control of the suitability of the quantities of pieces to be shipped. (Adapted from TVP Ltd, with permission.)

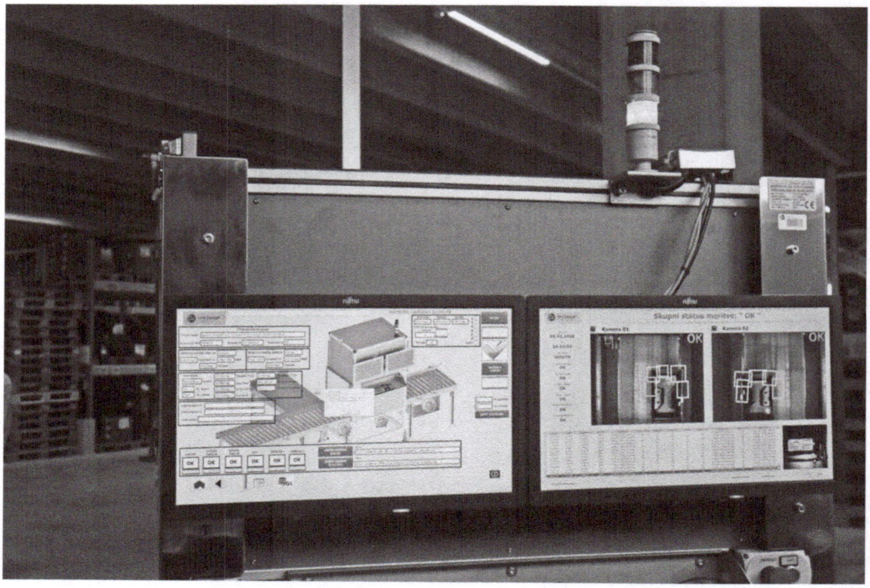

Figure 6.5 Autonomous production process from the input to the output of the production process. (Adapted from TVP Ltd, with permission.)

suitability of the quantities of pieces for shipment and Figure 6.5 shows the autonomous production process from the input to the output of the production process.

It is necessary to emphasise that changes in the direction of smart factories are already happening and will continue to happen even faster in the future. The fourth industrial revolution is going on imperceptibly and already demonstrates apparent effects on the most advanced companies and these companies build competitive advantage for future operations. Therefore, decisions and solutions in the field of automation of logistics in companies must be quick and comprehensive.

6.5 Discussion

6.5.1 Research specifics

The framework of the theoretical starting points, which represent the basic framework of our study, is based on the second-order cybernetics. The prefix "cyber" has become widespread since the early 1990s. It originates from the Wiener's (1948) coinage of "cybernetics" which derives from the Greek word "kubernetes", which in translation means the steersman. The significance could also be defined as the art of steering and steering the boat (Bai and Lindberg, 1999, p. 793). Despite its interdisciplinary, the term cybernetics has been used to describe the technical and engineering approach focused primarily on the study of information feedback loops, control systems and control intelligent machines (Mesjasz, 2006; Wiener, 1948).

The emergence of the so-called cybernetics of the first-order dates back to 40 years of the 20th century (van Dijkum and Schroots, 2006). During this period, the systemic approach was developing theoretically. Frank Fremont-Smith played an essential role in this process. Under Josiah Macy Jr. foundation, he took care of the promotion of the Macy Conferences between 1941 and 1960. Between scientists from various disciplines, meetings were held at these conferences in New York, with the explicit aim of promoting communication between different scientific disciplines and restoring unity in science. Conferences were the basis of the development of the fundamentals of system thinking and cybernetics (Dominici and Roblek, 2016, p. 226). After scientist set up, the foundations of first-order cybernetics framework beyond the control of feedback was developed.

For this reason, a large number of studies such as biological and open systems theory (von Bertalanffy, 1972), information and communication theory (Weaver and Shannon, 1964), self-organisation and regulation theory (Ashby, 1956), management and organisational cybernetics (Beer, 1985) were developed. In the 1980s of the 20th century, cybernetic science entered into the theory of management. Thus, Lynch and Kordis (1989) presented complexity theory as a new and essential paradigm for business development, for which Merry and Kassavin (1995) considered it to be a vital organisational tool. The development of information technologies in the 1970s led to the development of new channels of communication between different systems that are trying to steer each other. The second-order cybernetics was developed to solve problems in social studies (Geyer, 1995).

The second-order cybernetics is based on social and biological studies (von Foerster, 1984; Varela and Maturana, 1972). von Foerster (1984) defined cybernetics of the second-order as cybernetics of observation. The second-order cybernetic approach has bridged the micro-macro gap and led to closer integration of the individual with society (Bailey, 2006). Kieser and Kubicek (1992) were developed in the 1960s of the 20th century, a model for the contingency approach. This model helps to analyse the events in the business environment, which require an immediate response from managers, and it is expected that business decisions must be accepted quick and timely. Structural unpredictability is part of a behavioural theory of behaviour and systems that are designed to set up a particular business information system. The theory does not offer any single best solution. Efficient and effective management models cannot be easily adapted to unique environments. Each organisation locates specific features and functions in its environment. During this period, social cybernetics developed which encouraged the integration of the individual into society. It is a more subjective approach, in this case, more than a one-way transfer of knowledge (Bailey, 2006). In this context, socio-cybernetics also helps us to understand the phenomenon of networking and mutual communication in the human environment. Research focusing on the second-order cybernetic approach focuses on research problems such as instability, flexibility, learning changes, development and autonomy (Dominici and Roblek, 2016, p. 227).

In the context of a research of a particular system, we explore a living system that has its own "will", regardless of the level of development of organisms. A paradox is that living systems are organised organically and, on the other hand, information is open. Consequently, living systems are more challenging to manage. Their interactions with their environment are complicated, and it is impossible to predict more than a few steps in advance. Here the control capability, developed by second-order cybernetics, arises. The cybernetics is focused on the understanding of biological and social complexity as well as its control. For this reason, it is understandable that the cybernetics is focused on creating new forms (morphogenesis) and on a positive feedback loop. Despite the fact that first-order cybernetics involved influential biologists such as Bertalanffy (1975), one of the founders of the general system theory, the incentive for the development of cybernetics of the second-order came mainly from biology and neurophysiology, and later from epistemology in the framework of whose second-order cybernetics focuses on the very nature of knowledge, languages, cognition and communication.

6.5.2 *Framework of a smart factory complexity*

In the last few years, Industry 4.0 has become an integral part of the company's strategy. On the case of TPV Ltd., this is already evident from their vision statement, which is "Connecting innovative solutions". They are aware that innovation and knowledge are the cornerstones of successful development. That is why they want to use the potential of innovation, knowledge and ideas of other companies to create joint innovative, advanced and market-interesting solutions.

Based on their strategy, focused on developing the smart factory, TPV Ltd. has formed the quadruple helix embedding interdisciplinary integration of various partnership with the private as well as the public sector. They have a connection with the public institutes, universities faculties and inter-agency association. Within the private partnership, they rely on companies that are in the field of transformation in terms of Industry 4.0. Thus, the company pursue a purpose of the quadruple helix to encourage the perspective of the knowledge society and knowledge democracy for knowledge production and innovation (Carayannis, Barth, and Campbell, 2012).

Its contribution to the development of the company's internal environment is presented by the convergence between two until now unconnected worlds: (i) informational technology (IT) and (ii) operational technology (OT). Thus, technological developments at Industry 4.0 enable to establish a connection between the digital and the physical environment, thereby influencing fundamental organisational and process changes in the fields of supply chains, manufacturing processes, and even end-to-end ecosystems (Kusiak, 2018).

The cognitive computing based on artificial intelligence is leading companies to the new R&D which will soon result in transformations that will lead to complete self-managed organisation. Cognitive manufacturing transforms manufacturing in three focused ways: (i) intelligent assets and equipment, (ii) cognitive processes and operations and last that is the most important for our research question of (iii) smarter resources and optimisation which includes optimising and enhance resources such are labour and workforce (Mabkhot et al., 2018; Sjödin et al., 2018).

The companies' goal is integrating cognitive computing into Industry 4.0 solutions such as remotely steered, autonomy and robotisation moved from the IoTs vision. The cognitive computing based on artificial intelligence is leading companies to the new R&D, which will soon result in transformations that will lead to complete self-managed organisation. This can be achieved through machine intellect which enables the machine learning and their adoption according to the needs of the environment and tasks (Lee, Bagheri, and Kao, 2015). In a practical case, these mean that a machine (e.g. smart transport vehicle or machine vision systems) interacting (the machine is equipped with the RFID tag) with its social and machine networks to decide autonomously about production needs or product quality standard. In case of testing the quality, data-driven operational excellence which is based on the production information flow helps with the 3D machinery vision to detect the product defect and sort it out. This process is, hence, self-controlled. Figure 6.6 shows a smart factory module for 3D machine vision.

Question of ethics of responsibility is who will be responsible for the mistakes in the manufacturing processes and who will take over the responsibility for the products if the algorithms will blame a mistake that will have negative consequences for people.

Industry 4.0 brings to the organisation changes in organisational processes, working methods and staff structure, in particular, changes in their qualifications and competences. The smart factories are top-notch in terms of organisation. The key to these new factories includes crazy flexibility, openness and to look for people of this type. It is expected that the mastery mentality will be lost. Programmers will have tremendous competencies. Computer programming but systems work or do not work, game theory, the need for these staff will be almost endless.

Figure 6.6 Smart factory module for 3D piece tracking with machine vision for robotic insertion or removal of pieces. (Adapted from TVP Ltd, with permission.)

Macurová, Ludvík, and Žwaková (2017) point out that the high risk of organisations is the lack of skilled workers who will prepare, implement and use new technologies. The challenge in the future will, therefore, be job restructuring, as some of the less demanding occupations will disappear rapidly (Kane et al., 2015). Increasing productivity achieved through the use of smart technologies can help to provide jobs and increase consumer demand with additional income (countervailing effect), but the use of new manufacturing technologies and processes can also destroy jobs (the effects of redundancies). There is a concern that Industry 4.0 will lead to technological unemployment in the long run (Hungerland et al., 2015). Relatively, the competencies required in many jobs will change.

The fourth industrial revolution brings changes in organisational processes, working methods and staff structure, in particular, changes in their qualifications and competences. Increasing productivity achieved through the use of smart technologies can help to provide jobs and increase consumer demand with additional income (countervailing effect), but the use of new manufacturing technologies and processes can also destroy jobs (the effects of redundancies).

It is difficult to predict precisely how intensely the competencies that the production process staff will need in smart factories in the future, for example, by 2030, will change. The fact is that not all smart factories will be the same; they will also depend on the type of production. However, we should not overlook the fact that Industry 5.0 will come to life by 2030, which will be different from Industry 4.0. This industry is introducing so-called cooperative robots into production processes. These are robots that will be technologically able to work with humans in production processes. So, it will be about the connection between the machine and the human being. The essence of such cooperation will also be that human in the production processes will do every day work or work that is done by robots but will lead and rationalise all production processes. Industry 5.0 also means smaller factories, more manufacturing flexibility and personalisation of production as per customer requirements.

6.6 Conclusion

Development of the industry is going towards smart factories whose indispensable part is smart internal logistics. Currently, internal logistics processes are ineffective. In smart factories, process management will take place in real-time and consequently with maximum precision and efficiency. The goal is to increase the efficiency of business operations. Automation of internal logistics is undoubtedly an area of automation with great potential and the effect of improving efficiency. The presented solution has many advantages. It is optimal in terms of effect and investment; flexible, open and interconnected with other management systems and production automation. It is necessary to emphasise that changes in the direction of smart factories are already happening and will continue even faster in the future. The fourth industrial revolution is going on imperceptibly and already demonstrates apparent effects on the most advanced companies and these companies build competitive advantage for future operations. Therefore, decisions and solutions in the field of automation of logistics in companies must be quick and comprehensive.

In doing so, it is essential for the future of national economies, such as the Slovenian economy, that the links between companies and researchers exist. Young companies with great ideas, lots of energy and special skills, and larger, development-oriented companies with the necessary infrastructure and capital can jointly develop and successfully market innovative solutions that they would not be able to do on their own. Cooperation is an excellent potential for growth of the Slovenian and European Union economy.

References

Adler, P. 1983. Rethinking the skill requirements of new technologies. Working paper 9-784-027. Cambridge: Harvard University.

Allen, B.M., and A. Hirshon. 1998. Hanging together to avoid hanging separately: opportunities for academic libraries and consortia. Information technology and libraries 17, no. 1:36–44.

Ashby, W.R. 1956. The effect of experience on a determinate dynamic system. Behavioral science 1, no. 1:35–42.

Attewell, P. 1985. The automated office: a case study. New York: Stony Brook.
Attewell. P., and J. Rule. 1984. Computing and organizations: what we know and what we don't know. Communications of the ACM 27, no. 12:1184–1192.
Bai, G., and Å.L. Lindberg. 1999. A sociocybernetic approach to information systems development. Kybernetes 28, no. 6/7:792–809.
Bailey, K.D. 2006. Sociocybernetics and social entropy theory. Kybernetes 35, no. 3/4:375–384.
Beer, S. 1985. Diagnosing the system for organizations. Chichester: John Wiley & Sons.
Bekkers, M.J.J.V., and P.H.A. Frissen. 1992. Informalization and administrative modernization: a comparative analysis. In European public administration and informatization: a comparative research project into policies, systems, infrastructures and projects, ed. P.H.A. Frissen, V.J.J.M. Bekkers, Brussaard, I.T.M. Snellen, and M. Wolters, 479–490. Amsterdam: IOS Press.
Bellamy, C., and J.A. Taylor. 1998. Governing in the information age. Buckingham: Open University Press.
Bisson, P., E. Stephenson, and P.S. Vigurie. 2010. The global grid. McKinsey quarterly. https://www.mckinsey.com/business-functions/strategy-and-corporate-finance/our-insights/the-global-grid (accessed June 10, 2019).
Braverman, H. 1974. The degradation of work in the twentieth century. New York: Monthly Review Press.
Braunerhjelm, P., and A. Henrekson. 2016. An innovation policy framework: bridging the gap between industrial dynamics and growth. In Essays in public sector entrepreneurship, ed. D.B. Audretsch, and A.N. Link, 95–130. Cham: Springer.
Cambridge Dictionary, s.v. "Virtualisation". https://dictionary.cambridge.org/dictionary/english/virtualization (accessed September 19, 2019).
Carayannis, G.E., T.D. Barth, and D.F.J. Campbell. 2012. The Quintuple Helix innovation model: global warming as a challenge and driver for innovation. Journal of innovation and entrepreneurship 1, no. 2. https://innovation-entrepreneurship.springeropen.com/articles/10.1186/2192-5372-1-2 (accessed September 3, 2019).
Castells, M. 2010. The rise of the network society: the information age – economy, society, and culture, vol. 1. Chichester: John Wiley & Sons.
Ceruzzi, P.E. 2003. A history of modern computing. Cambridge: MIT Press.
Chow, S.K., Y.Q. Chin, Y.H. Lee, C.H. Leung, and H.F. Tang. 2012. Nurses' perceptions and attitudes towards computerisation in a private hospital. Journal of clinical nursing 21, no. 11–12:1685–1696.
Coe, C.K., and F. Borthick. 1984. Computer auditing in the federal government. Public budgeting & finance 4, no. 3:67–74.
Cooper, J., and A. James. 2009. Challenges for database management in the internet of things. IETE technical review 26:320–329.
Davis, F.D., P.R. Bagozzi, and R.P. Warshaw. 1989. User acceptance of computer technology: a comparison of two theoretical models. Management science 35, no.8:982–1003.
Dominici, G., and V. Roblek. 2016. Complexity theory for a new managerial paradigm: a research framework. In Neostrategic management: contributions to management science, ed. I. Vrdoljak Ragus, N. Podrug, and L. Jelenc, 223–241. Chem: Springer International Publishing AG.

Donate, M.J., and S.D.J. de Pablo. 2015. The role of knowledge-oriented leadership in knowledge management practices and innovation. Journal of business research 68, no. 2:360–370.
Easterby-Smith, M., R., Thorpe, R.P. Jackson. 2012. Management research, 4th ed. London: Sage.
Etzkowitz, H., and L. Leydesdorff. 2000. The dynamics of innovation: from national systems and "mode 2" to a triple helix of university–industry–government relations. Research policy 29, no. 2:109–123.
Geyer, F. 1995. The challenge of sociocybernetics. Kybernetes 24, no. 4:6–32.
Glenn, E.N., and L.R. Feldberg. 1977. Degraded and deskilled: the proletarianization of clerical work. Social problems 25, no. 1:52–64.
Giuliano, V.E. 1982. The mechanization of office work. Scientific American 247, no. 3:148–165.
Gregory, J., and K. Nussbaum. 1982. Race against time: automation of the office – an analysis of the trends in office automation and the impact on the office workforce. Office technology and people 1, no. 2/3:197–236.
Gruber F.E. 2013. Industry 4.0: a best practice project of the automotive industry. In Digital product and process development Systems. IFIP advances in information and communication technology, eds., L.G. Kovács, and D. Kochan, 36–40. Heidelberg: Springer.
Guchelaar, H.J., and M.D. Kalmeijer. 2003. The potential role of computerisation and information technology in improving prescribing in hospitals. Pharmacy world and science 25, no. 3:83–87.
Guile, D. 2010. The learning challenge of the knowledge economy. Rotterdam: Sense.
Hanna, N.K. 2016. Mastering digital transformation: towards a smarter society, economy, city and nation. London: Emerald Group Publishing Limited.
Hirschhorn, L. 1984. Beyond mechanization: work and technology in a post-industrial age. Boston: MIT Press.
Hungerland, F., J. Quitzau, C. Zuber, L. Ehrlich, C. Growitsch, C.M. Rische, and J.H. Haß. 2015. The digital economy (No. 21e). Strategy 2030—wealth and life in the next generation. http://www.econstor.eu/handle/10419/121322 (accessed September 12, 2019).
Kane, G.C., D. Palmer, N.A. Phillips, and D. Kiron. 2015. Is your business ready for a digital future? MIT sloan management review 56(Summer):37–44. http://ilp.mit.edu/media/news_articles/smr/2015/56415.pdf (accessed September 12, 2019).
Kaplan, A.M., and M. Haenlein. 2010. Users of the world, unite! The challenges and opportunities of social media. Business horizons 53, no. 1:59–68.
Kaplan, R.S., and A. Mikes. 2012. Managing risks: a new framework. Harvard business review 90, no. 6:48–60.
Katsikas, S.K., and S. Gritzalis. 2017. Digitalization in Greece: state of play, barriers, challenges, solutions. In Beyond bureaucracy, ed. A.A. Paulin, L.G. Anthopoulos, and C.G. Reddick, 355–375. Cham: Springer.
Kelly, P.T.F. 1978. Public packet switched data networks, international plans and standards. Proceedings of the IEEE 66, no. 11:1539–1549.
Kieser, A., and H. Kubicek. 1992. Organisation. Berlin: W. de Gruyter.
Kusiak, A. 2018. Smart manufacturing. International journal of production research 56, nos. 1–2:508–517.

Kuula, M., A. Putkiranta, and J. Toivanen. 2012. Coping with the change: a longitudinal study into the changing manufacturing practices. International journal of operations & production management 32, no. 2:106–120.
Lasi, H., P. Fettke, G.H. Kemper, T. Feld, and M. Hoffmann. 2014. Industry 4.0. Business & information systems engineering 6:239–242.
Lee, J., B. Bagheri, and A.H. Kao. 2015. A cyber-physical systems architecture for Industry 4.0-based manufacturing systems. Manufacturing letters 3:18–23.
Lekše, I., J. Sluga, and D. Rajšelj. 2019a. Naredimo kompleksnost enostavno. Internal papers.
Lekše, I., J, Sluga, and D. Rajšelj. 2019b. Automation of internal logistics with automated guided vehicles – AGVs. Internal papers.
Lynch, D., and L.P. Kordis. 1989. Strategy of the dolphin: scoring a win in a chaotic world. New York: Fawcett Columbine.
Mabkhot, M.M., M.A. Al-Ahmari, B. Salah, and H. Alkhalefah. 2018. Requirements of the smart factory system: a survey and perspective. Machines 6, no. 2(June):1–22. https://www.mdpi.com/2075-1702/6/2/23.
Macurová, P., L. Ludvík, and M. Žwaková. 2017. The driving factors, risks and barriers of the industry 4.0 concept. https://www.researchgate.net/publication/323238209_The_driving_factors_risks_and_barriers_of_the_industry_40_concept (accessed September 11, 2019).
Merriam-Webster Dictionary, s.v. "digitalisation". https://www.merriam-webster.com/dictionary/digitalization (accessed May 14, 2019).
Merry, U., and N. Kassavin. 1995. Coping with uncertainty: insights from the new sciences of chaos, self-organization, and complexity. Westport: Praeger Publishers/Greenwood Publishing Group.
Mesjasz, C. 2006. Przegląd teoretycznych koncepcji nadzoru korporacyjnego. Zeszyty Naukowe/Akademia Ekonomiczna w Krakowie, 700:47–62.
Mosconi, F. 2015. The new European industrial policy: global competitiveness and the manufacturing renaissance. London: Routledge.
Murphree. M.C. 1984. Brave new office: the changing world of the legal secretary. In My troubles are going to have trouble with me: everyday trials and triumphs of women workers, ed. K. Sacks, and D. Remy, 140–159. New Brunswick: Rutgers University Press.
Ning, H., and H. Liu. 2015. Cyber-physical-social-thinking space based science and technology framework for the Internet of things. Science china information sciences 58:1–19.
Oesterreich, T.D., and F. Teuteberg. 2016. Understanding the implications of digitisation and automation in the context of Industry 4.0: a triangulation approach and elements of a research agenda for the construction industry. Computer industry 83:121–139.
Patton, M.Q. 2002. Two decades of developments in qualitative inquiry: a personal, experiential perspective. Qualitative social work 1, no. 3:261–283.
Paulin, A. 2019. Smart city governance. Amsterdam: Elsevier.
Pélissié du Rausas, M., J. Manyika, E. Hazan, J. Bughin, M. Chui, and R. Said. (2011). Internet matters: the net's sweeping impact on growth, jobs and prosperity. New York: McKinsey Global Institute. https://www.mckinsey.com/~/media/McKinsey/Industries/High%20Tech/Our%20Insights/Internet%20matters/MGI_internet_matters_exec_summary.ashx (accessed August 18, 2019).

Poppel, H.L. 1982. Who needs the office of the future? Harvard business review 60, no. 6:146–155.
Porter, M. 2001. Strategy and the internet. Harvard business review 79, no.3:63–78.
Roblek, V., I. Erenda, and M. Meško. 2018. The challenges of sustainable business development in the post-industrial society in the first half of the 21st century. In Managerial strategies for business sustainability during turbulent times, ed. R.D. Leon, pp. 1–22. Hersley: IGI Global.
Roblek, V., M. Meško, and A. Krapež. 2016. A complex view of industry 4.0. Sage open 6, no. 2:1–12.
Roblek, V., M. Meško, M. Pejić Bach, and A. Bertoncelj. 2014. "Impact of knowledge management on sustainable development in the innovative economy." In Advancements in economic and managerial theory and practice: business systems laboratory – 2nd international symposium systems thinking for a sustainable economy, Rome, 23–24 January 2014, 11–24. Rome: Universitas Mercatorum.
Roblek, V., M. Pejić Bach, M. Meško, and A. Bertoncelj. 2013. The impact of social media to value added in knowledge-based industries. Kybernetes 42, no. 4:554–568.
Schaupp, S., and R. Diab. 2020. From the smart factory to the self-organisation of capital:'Industrie 4.0'as the cybernetisation of production. Ephemera 20, no. 2 (In press). http://www.ephemerajournal.org/contribution/smart-factory-self-organisation-capital-'industrie-40'-cybernetisation-production.
Schiepek, G. 1999. Selbstorganisation in psychischen und sozialen prozessen: neue perspektiven der psychotherapie, ed. K. Mainzer, Komplexe systeme und nichtlineare dynamik in natur und gesellschaft. https://link.springer.com/book/10.1007/978-3-642-60063-0 (accessed June 8, 2019).
Schutt, R.K. 2012. Investigating the social world. 7 ed. Thousand Oaks, CA: SAGE Publications.
Schwab, K. 2016. The fourth industrial revolution: what it means, how to respond. World Economic Forum. https://www.weforum.org/agenda/2016/01/the-fourth-industrial-revolution-what-it-means-and-how-to-respond (accessed August 4, 2019).
Shaiken, H. 1988. High tech goes 3rd-world. Technology review 91, no. 1:38.
Sjödin, D.R., V. Parida, M. Leksell, and A. Petrovic. 2018. Smart factory implementation and process innovation: a preliminary maturity model for leveraging digitalization in manufacturing moving to smart factories presents specific challenges that can be addressed through a structured approach focused on people, processes, and technologies. Research-technology management 61, no. 5:22–31.
Ślusarczyk, B. 2018. Industry 4.0: are we ready? Polish journal of management studies 17, no. 1:232–247.
Taalbi, J. 2018. Origins and pathways of innovation in the third industrial revolution. Industrial and corporate change 28, no. 5:1125–1148.
Turing, A. 2018. How can we design fair, transparent, and accountable AI and robotics? https://www.turing.ac.uk/blog/can-design-fair-transparent-accountable-ai-robotics/ (accessed August 19, 2019).
Turner, J. 1984. Computer mediated work: the interplay between technology and structured jobs-claims representatives in the social security administration. Working paper, CRIS 80. New York: Graduate School of Business Administration, New York University.

van Dijkum, C., and J. JF Schroots. 2006. The challenge of the past for the future of the social sciences. Kybernetes 35, no. 3/4:385–402.
Varela, F., and H. Maturana. 1972. Mechanism and biological explanation. Philosophy of science 39, no. 3:378–382.
von Bertalanffy, L. 1972. The history and status of general systems theory. Academy of management journal 15, no. 4:407–426.
von Bertalanffy, L. 1975. Perspectives on general system theory. New York: George Braziller.
von Foerster, H. 1984. Observing systems. Seaside: Intersystems Publications.
Weaver, W., and C. Shannon. 1964. The mathematical theory of information. Urbana, IL: Illinois Press.
Wiener, N. 1948. Cybernetics or control and communication in the animal and the machine. Cambridge, MA: MIT Press.
Yin, R.K. 2009. Case study research: design and methods, 4th ed. Thousand Oaks, CA: Sage.
Yin, R.K., and D. Davis. 2007. Adding new dimensions to case study evaluations: the case of evaluating comprehensive reforms. New directions for evaluation 113:75–93.
Zuboff, S. 1985. Automate-informate: the new faces of intelligent technology. Organizational dynamic 14:5–18.
Zuboff, S. 1988. In the age of the smart machine. New York: Basic Books.
Zuboff, S., ed. 2019. The age of surveillance capitalism: the fight for a human future at the new frontier of power. London: Profile Books Ltd.

chapter seven

Maturity level assessment for industry 4.0 integration into Lean Manufacturing

Marcos Malinverni Pagliosa, Guilherme Luz Tortorella, and João Carlos Espíndola Ferreira
Universidade Federal de Santa Catarina, Florianópolis, Brazil

Contents

7.1 Introduction

The term Industry 4.0 (I4.0) was coined in 2011 during the Hannover fair in Germany and refers to the new technological revolution that the industry is going through. I4.0 represents a new phase in industrial evolution and offers significant technological advances that allow organizations to operate at the highest levels of quality, cost, and performance (Kagermann et al., 2013). The technologies inherent to I4.0 promote autonomous and dynamic production, embracing the concept of smart manufacturing and enabling value chain members to intensively exchange data in real time. In addition, these technologies provide a modular and highly customized mass production system that meets market's requirements on flexibility, agility, and low cost (Radziwon et al., 2014; Kolberg et al., 2017).

The emergence of new disruptive concepts and technologies is part of the dynamic and extremely competitive environment in which manufacturing companies are inserted (Wilding et al., 2012). Due to the range of benefits that I4.0 can offer, many organizations intend to invest in acquiring such technologies. According to a survey conducted by the National Confederation of Industry (CNI Brazil, 2016), 48% of the large Brazilian companies plan to invest in I4.0, and the main interest is in technologies aimed at increasing the efficiency and performance of the production process.

However, the existence of strategies that promote greater competitiveness of organizations precedes I4.0. Among the most commonly adopted approaches, it is worth highlighting the Lean Manufacturing (Lean). Lean is characterized as a socio-technical approach that encompasses organizational practices and principles aiming at systematically eliminating waste, improving productivity and quality with a strong focus on customer requirements (Womack and Jones, 2004; Junior and Filho, 2010; Jasti and Kodali, 2015). The simplicity of its application and the results achieved are some of the reasons why Lean has been widely adopted in a large number of organizations (Simpson and Power, 2005; Dennis, 2008; Holm and Ahlstrom; 2010; Burch et al., 2019).

However, Lean, is usually considered a systematic human-centered approach that stands out for its simplicity and effectiveness (Dickmann, 2008); not necessarily requiring the application of high technologies (Seppälä and Klemola, 2004). Such aspect may present conflicting points with the advanced technologies included in I4.0. According to Tortorella et al. (2018), the concurrent implementation of I4.0 technologies and lean practices (LPs) in manufacturing companies may raise contradictory arguments given their different characteristics. In addition, as many organizations still struggle to understand the general idea of I4.0 (Erol et al., 2016), its specific concepts and principles (Sanders et al., 2016) may compromise the analysis of possible impacts from its adoption. Thus, despite the growing awareness that this topic has gained, there are questions about the feasibility and effectiveness of integrating I4.0 and its technologies into Lean and its LPs on important aspects of manufacturing.

Based on the relevance of both I4.0 and Lean for the industrial context and the scarcity of studies that examine its effects (Gjeldum et al., 2016; Landscheidt and Kans, 2016; Kolberg et al., 2017), the following research question can be raised: "*how to identify which productive areas are most prone to the concomitant implementation of I4.0 and Lean?*" To answer this question, this research aims to propose an assessment method that allows prioritizing the implementation of I4.0 and Lean in productive areas of manufacturing companies based on their maturity level. An exploratory research was conducted in a large Brazilian auto parts manufacturer, which has been implementing Lean for more than 10 years and presents initiatives of I4.0 implementation. The proposed method is composed of procedures that involve both qualitative and quantitative analysis of the context under study. Besides the theoretical contribution already highlighted, this research has practical implications. It offers guidelines to assist the implementation of I4.0 and Lean, providing guidelines for a more assertive implementation in the organization's context.

The structure of this work is subdivided as follows. Sections 1.1 and 1.2 present a literature review on I4.0 and Lean, respectively. Section 2 describes the proposed method, whose results are displayed in Section 3. Finally, Section 4 concludes the paper and indicates future research directions.

7.2 *Industry 4.0*

Deemed as the fourth industrial revolution, I4.0 represents a set of technologies and concepts that can be applied to the value chain of organizations through the use of information and communication technologies, such as cyber-physical systems (CPSs), Internet of Things (IoT), and Big Data. The integration of I4.0 has the potential to significantly change existing value chains, their business models, and processes (Schmidt et al., 2015). In addition, it promotes digital integration that enables value chain members to intensively exchange data and information in real time, creating possibilities for process improvement (Tamás et al., 2016; Gubán and Kovács, 2017; Sanders et al., 2017). According to Kagermann et al. (2013), the use of I4.0 technologies can increase product quality while making manufacturing processes more efficient. It allows manufacturing system to be modular and flexible, thus enabling mass customization (Lasi et al., 2014; Posada et al., 2015; CNI Brazil, 2016; Tamás et al., 2016; Xu and Chen, 2016).

Despite the growing emphasis that the theme has received, several organizations still have difficulty understanding the general idea, principles, and concepts of I4.0 (Sanders et al., 2016). To facilitate understanding of this topic, minimize errors and misinterpretations, and guide the implementation of I4.0, some roadmaps and maturity models have been proposed in the literature (e.g., Anderl, 2014; Lee et al., 2015; Chukwuekwe et al., 2016; Leyh et al., 2016, 2017; Tonelli et al., 2016; Schuh et al., 2017; Ghobakhloo, 2018). Ganzarain and Errasti (2016) formulated a maturity model, consisting of three stages and five levels that help companies create a customized roadmap, oriented to I4.0. Pokorni et al. (2017) developed a model with 33 criteria and 130 indicators, designed to evaluate the maturity level of both I4.0 and Lean in manufacturing environments. Price Waterhouse Coopers (PWC) has structured a seven-dimensional four-stage maturity model for organizations that are beginning the I4.0 adoption (Geissbauer, 2016). The German government has proposed a maturity model for evaluating I4.0 with 62 items grouped into nine dimensions namely: strategy, leadership, customers, products, operations, culture, people, governance, and technology (Schumacher et al., 2016).

However, despite these initiatives, it is clear that there is still much to understand and investigate with regards to the impacts that the I4.0 adoption can have on manufacturing organizations (Gjeldum et al., 2016; Kolberg et al., 2016; Landscheidt and Kans, 2016; Xu and Chen, 2016; Sanders et al., 2017; Santorella, 2017). Regarding the main technologies of I4.0, several researchers have addressed them in the literature, proposing different classifications. Pagliosa et al. (2018) conducted a study based on 93 articles to identify nine main I4.0 technologies related with the manufacturing context; they are as follows: IoT, CPS, Cloud computing, Big Data, horizontal/vertical integration, advanced robots, augmented reality, simulation, and additive manufacturing (see Table 7.1).

Regarding the maturity models for I4.0, literature also presents a variety of research and approaches for this purpose (Anderl, 2014; Lee et al., 2015; Chukwuekwe et al., 2016; Ganzarain and Errasti, 2016; Leyh et al., 2016; Geissbauer, 2016; Schumacher et al., 2016; Tonelli et al., 2016; Leyh et al., 2017; Pokorni et al., 2017; Schuh et al., 2017; Ghobakhloo, 2018). However, only a portion of this evidence includes manufacturing environments. In view of this, Table 7.2 consolidates the main dimensions and criteria used in some of the maturity models developed within such context. Fifty-five criteria were critically analyzed and redundancies eliminated, resulting in 20 main criteria for assessing the maturity level (i) for I4.0 implementation. These were then grouped by similarity into four main dimensions, they are as follows: (di_1) vision of the future, (di_2) culture, (di_3) operations, and (di_4) technology and data analysis. Dimension di_1 refers to organization's ability to understand the context of I4.0, builds strategic planning aligned with that context, and provides the necessary resources. Dimension di_2 represents the initiatives aimed at the development of an organizational culture coherent with I4.0. di_3 comprises the activities focused on important aspects of I4.0, such as: decentralization, virtualization, real-time operations, and analysis. Finally, di_4 refers to aspects with a

Table 7.1 Main Industry 4.0 technologies

Code	I4.0 technologies	Some authors
t_1	Internet of Things (IoT)	Züehlke (2010), Tao et al. (2011), Spath et al. (2013), Fleisch et al. (2015), Arnold (2016), Alias et al. (2017)
t_2	Cyber-physical system (CPS)	Broy et al. (2012), Spath et al. (2013), Brettel et al. (2014), Kern et al. (2015), Chiang and Lee (2017)
t_3	Cloud computing	Spath et al. (2013), Ren et al. (2015), Rüßmann et al. (2015), Stock and Seliger (2016), Marjani et al. (2017), Mora et al. (2017), Sackey et al. (2017)
t_4	Big Data	Lee et al. (2015), Bagheri et al. (2015), Sanders et al. (2016), Trstenjak and Cosic (2017), Wagner et al. (2017)
t_5	Horizontal/vertical integration	Tao et al. (2011), Struker and Weppner (2012), Spath et al. (2013), Kibira et al. (2015), Chen et al. (2017)
t_6	Advanced robots	Almada-Lobo (2016), Edwards and Ramirez (2016), Erol et al. (2016), Powell et al. (2013) Mourtzis et al. (2017), Sackey et al. (2017), Strandhagen et al. (2017)
t_7	Augmented reality	Yoon et al. (2012), Kolberg and Züehlke (2015), Ranz et al. (2015), Arnold (2016), Liao et al. (2017)
t_8	Simulation	Schuh et al. (2017), Kibira et al. (2015), Gronau (2016), Sanders et al. (2016), Tamás (2017)
t_9	Additive manufacturing	Dugenske and louchez (2014), Schumacher et al. (2016), Stock and Seliger (2016), Powell et al. (2013)

Source: Pagliosa et al. (2018).

strong technological nature that allow for intense and safe horizontal and vertical digitization throughout the company.

7.3 Lean Manufacturing

Lean can be defined as a productive approach that aims to reduce waste along the value chain, improving productivity and quality from a customer perspective (Ohno, 1988; Shah and Ward, 2003; Womack and Jones, 2004; Junior and Filho, 2010; Liker and Franz, 2011; Jasti and Kodali, 2015). Lean aims to reduce the variation in production processes, thereby offering performance improvements and cost savings (Ohno, 1988; Womack and Jones, 2004; Dennis, 2008). This approach stands out for its simplicity and effectiveness, reinforcing a strategic alignment between the needs of the business and people (Simpson and Power, 2005; Dennis, 2008; Burch et al., 2019).

Table 7.2 Dimensions and criteria for maturity in I4.0

Dimensions	I4.0 maturity assessment criteria by author					Consolidation of maturity criteria
	Ganzarain and Errasti (2016)	Leyh et al. (2016)	Geissbauer (2016)	Schumacher et al. (2016)	Schuh et al. (2017)	
di_1 view of future	I4.0 projects		Disruptive business models and focus on core competency	Adaption of new business models		i_1 New business plan
	Understanding of I4.0	I4.0 works as a facilitator of company strategy				i_2 Understanding of I4.0 context
	Capacity and resources analysis			Available resources for new business models	Continuous, agile adaptation to a changing environment	i_3 Available resources
di_2 culture		Actively engaged with I4.0 topics	Collaboration as a key value driver	Open innovation and cross-company collaboration	Understanding the context and produce knowledge	i_4 Knowledge sharing
		Collaborates strongly with its business partners			Collaboration within the value network	i_5 Collaboration
		Flexible culture	Innovative concepts and technologies	Openness of employees to new technology		i_6 Flexible culture
	Training capacitation		I4.0 culture and training	ICT[a] competences of employees	Training programs	i_7 I4.0 training
				Willingness of leaders	More flexible approach to decision-making	i_8 I4.0 leadership
			Impact analysis and decision support	Management competences and methods		i_9 Decision support
	Risk management			Central coordination for I40		i_{10} I4.0 coordination

Table 7.2 (Continued) Dimensions and criteria for maturity in I4.0

	I4.0 maturity assessment criteria by author					
Dimensions	**Ganzarain and Errasti (2016)**	**Leyh et al. (2016)**	**Geissbauer (2016)**	**Schumacher et al. (2016)**	**Schuh et al. (2017)**	**Consolidation of maturity criteria**
di_3 operations			Decentralized autonomy	Decentralization of processes	Delegate decisions to ICT	i_{11} Decentralization
			Virtualized processes	Modeling and simulation	Predictive capacity	i_{12} Virtualization
	Capacity and resources analysis		Predictive analytics			i_{13} Capacity analysis
			Real-time optimization		Automatic monitoring and real-time documentation	i_{14} Real-time operations
di_4 technology and data analysis	Identify requirements	Automated information flows	Secure data exchange	Moderns ICT	ICT security	i_{15} ICT and data security
		Optimization of value networks	External data integration functionalities		Information exchange between companies	
	Identify technology involved in I4.0		Intelligent database	Utilization of mobile devices	Continuous process improvement based on database	i_{16} I4.0 technologies
		Integration of machines and systems		Utilization of machine-to-machine communication		i_{17} Advanced machines
			Flexible organizations		Capable agile adaptation to a changing environment	i_{18} Flexibility
			Self-learning algorithm			i_{19} Artificial intelligence
		Wide complete internal/enterprise integration of all systems	Fully digitized, integrated partner ecosystem	Decentralization of processes	Digitalization	i_{20} Digitalization

According to Hines et al. (2004), the understanding of Lean and its practices have been evolving as companies advance in their implementation. Maturity models are a useful tool that enables companies to access the level of their organizational processes (Albliwi et al., 2014). Literature contains a large number of works on Lean and maturity models (Hamdi and Leite, 2012; Powell et al., 2013; Verrier et al., 2016). However, according to Tortorella and Fettermann (2018), there is no consensus on a single way to measure the degree of maturity of Lean. Shah and Ward (2007) proposed a structure composed of 10 bundles and 41 LPs, empirically validating it. In this sense, approaches based on the adoption levels of some predefined LPs has been used (Netland and Ferdows, 2014; Marodin et al., 2015). Additionally, several studies have shown the positive association between the implementation of LPs and the improvement of operational performance (Conger and Miller, 2013; Kanigolla et al., 2014; Rezende et al., 2016). More specifically, some studies report the integration between I4.0 technologies into some LPs, which are listed in Table 7.3.

Regarding the maturity models for Lean, Table 7.4 presents some of the main models proposed in the literature, which total 103 criteria. These were critically analyzed, and their redundancies were disregarded, leading to 34 main criteria for the evaluation of the Lean implementation maturity (l). These were then grouped according to similarity into eight dimensions, namely (dl_1) forward thinking, (dl_2) culture, (dl_3) operations, (dl_4) quality, (dl_5) customer focus, (dl_6) costs, (dl_7) suppliers, and (dl_8) inventory. The dl_1 dimension represents the organization's ability to anticipate short-, medium-, and long-term changes. Dimension dl_2 aims at employees' performance evaluation, recognition, and development. dl_3 addresses the set of key practices that promote a better operational management, such as TPM (total productive maintenance) and SMED. Dimension dl_4 refers to organization's ability to manage quality performance, mitigate costs, and prevent failures. In turn, dl_5 is focused on adding value from the eyes of the customer in aspects such as customization, delivery, and relationship management. The dl_6 dimension covers the management of direct and indirect costs related to the company's activities. dl_7 includes activities related to partnerships, performance evaluation, and supplier involvement. Finally, dl_8 represents activities related to the management of different types of inventory and production leveling.

7.4 Proposed methodology

To reach the objective proposed by this work, our method comprises procedures that involve both qualitative and quantitative analysis of the context under study. The method consists of four main steps: (i) select the evaluation criteria and productive areas of the company, (ii) evaluate the dimensions important for I4.0 and Lean, (iii) assess the development level of I4.0 and Lean criteria in the productive areas, and (iv) rank and prioritize areas for a concomitant implementation of I4.0 and Lean.

Table 7.3 Main Lean practices

Code	Lean practices	Some authors
p_1	Kanban	Kouri et al. (2008), Dave et al. (2016), Kolberg et al. (2017), Tortorella and Fettermann (2018)
p_2	Value stream mapping (VSM)	Bhamu and Singh (2014), Villalba et al. (2015), Tamás et al. (2016), Ferrera et al. (2017)
p_3	Poka-yoke	Kolberg and Züehlke (2015), Mrugalska and Wyrwicka (2017)
p_4	Kaizen	Bhamu and Singh (2014), Martinez et al. (2016), Sanders et al. (2017), Dombrowski et al. (2017)
	Pull production	Dombrowski et al. (2017), Ferrera et al. (2017), Kolberg et al. (2017)
p_6	Andon	Bhamu and Singh (2014), Dave et al. (2016), Mrugalska and Wyrwicka (2017)
p_7	SMED – Single Minute Exchange of Die	Martinez et al. (2016), Sanders et al. (2016), Kusters et al. (2017), Sanders et al. (2017)
p_8	Standardization	Kouri et al. (2010), Dombrowski et al. (2017), Illés et al. (2017), Kolberg et al. (2017)
p_9	Total quality management (TQM)	Ibira et al. (2015), Eleftheriadis and Myklebust (2016), Illés et al. (2017), Dombrowski et al. (2017)
p_{10}	Takt time	Kouri et al. (2010), Bhamu and Singh (2014), Salmi et al. (2015), Aydos and Ferreira (2016)
p_{11}	Heijunka	Kouri et al. (2010), Bhamu and Singh (2014), Tamás et al. (2016), Kolberg et al. (2017)
p_{12}	5S	Bhamu and Singh (2014), Tamás et al. (2016), Illés et al. (2017), Sanders et al. (2017)
p_{13}	Jidoka	Bhamu and Singh (2014), Sanders et al. (2016), Tamás et al. (2016)
p_{14}	Total productive maintenance (TPM)	Bhamu and Singh (2014), Sanders et al. (2016, 2017)

Source: Pagliosa et al. (2018).

Table 7.4 Dimensions and maturity criteria for lean

Dimensions	Lean maturity assessment criteria by author: Doolen and Hacker (2005)	Bhasin (2008)	Pakdil and Leonard (2014)	Maasouman and Demirli (2015)	Santos and Tontini (2018)	Consolidation of maturity criteria	
dl_1 view of future		Anticipating future changes			Strategic planning based on long-term philosophy	l_1	Strategic planning
		New market and new technology development			Business plan	l_2	Business plan
dl_2 culture	Formal reward system	Labor productivity	Total of employees involved in lean practices/total employees	Empowerment, improvement, and motivation	Programs of involvement and recognition	l_3	Programs of involvement and recognition
	Multifunctional workforce			Team work	Multifunctional teams	l_4	Team work
	Employee evaluation	Quality of leadership development	Total of employees working in teams/total employees	Leadership performance and goal setting and action planning	Leadership development	l_5	Leadership performance
		Quality of professional/technical development			Training	l_6	Training programs

Table 7.4 (Continued) Dimensions and maturity criteria for lean

Dimensions	Lean maturity assessment criteria by author					Consolidation of maturity criteria	
	Doolen and Hacker (2005)	Bhasin (2008)	Pakdil and Leonard (2014)	Maasouman and Demirli (2015)	Santos and Tontini (2018)		
		Absenteeism	Absenteeism rate	Absenteeism rate		l_7	Absenteeism
		Labor turnover	Labor turnover rate			l_8	Turnover
		Retention of top employees					
dl_3 operations	Pay for performance	Labor productivity				l_9	Operational performance
				Standardization of production processes	Processes and tools	l_{10}	Standardization of processes
	Work standardization		Takt time	Standardization of daily maintenance activities			
	Setup time reduction		Average set-up time		Setups	l_{11}	Setups
	Cycle time reduction	Shorter cycle time	OEE[a]	OEE		l_{12}	OEE

(Continued)

Table 7.4 (Continued) Dimensions and maturity criteria for lean

Dimensions	Lean maturity assessment criteria by author					Consolidation of maturity criteria
	Doolen and Hacker (2005)	Bhasin (2008)	Pakdil and Leonard (2014)	Maasouman and Demirli (2015)	Santos and Tontini (2018)	
	TPM	Capital efficiency	Total time spent on unplanned or emergency repairs/total maintenance time	Preventive, predictive, and proactive maintenance	TPM	l_{13} TPM l_{14} Preventive maintenance
		Health and safety per employee		Safety, ergonomics and environmental		l_{15} Safety
	Pull flow				Pull system	l_{16} Pull system
			Total of problem-solving teams/total employees	Process capability analysis and improvement	Continuous improvement	l_{17} Continuous improvement
				Analysis and problem-solving processes	Problem solving, people	

Table 7.4 (Continued) Dimensions and maturity criteria for lean

Dimensions	Lean maturity assessment criteria by author					Consolidation of maturity criteria
	Doolen and Hacker (2005)	Bhasin (2008)	Pakdil and Leonard (2014)	Maasouman and Demirli (2015)	Santos and Tontini (2018)	
dl_4 quality	Total quality management (TQM)	Quality ratings	Total of people dedicated to quality control/total employees		Quality at source	l_{18} Quality management
		Defects of critical products/components			Product development that considers the needs of external customers	
		Quality of new product development and project management processes	Rework rate	Scrap and rework rate	Quality zero defect	l_{19} Scrap and rework
		Quality costs	Total cost of poor quality/total costs			l_{20} Quality costs
	Mistake or error proofing		Total scraps/total sales	Preventive quality control (FMEA, SPC[b], Poka Yoke)		l_{21} Mistake proofing

(Continued)

Table 7.4 (Continued) Dimensions and maturity criteria for lean

Dimensions	Lean maturity assessment criteria by author					Consolidation of maturity criteria	
	Doolen and Hacker (2005)	Bhasin (2008)	Pakdil and Leonard (2014)	Maasouman and Demirli (2015)	Santos and Tontini (2018)		
dl_5 focus on the customer	Product customization	Time to market for new products			Manages customer relationships effectively	l_{22}	Manages customer relationships
		Customer satisfaction index					
	Customer requirements analysis	Customer retention rate	Customer satisfaction index		Customer satisfaction	l_{23}	Customer satisfaction
		Responsiveness (customer defined)	Customer complaint rate				
	Services to enhance value				Customer-oriented quality	l_{24}	Customer oriented
	Delivery performance	On time delivery (customer defined)	Customer retention rate	On time delivery		l_{25}	On time delivery
dl_6 costs	Total cost evaluation	Manufacturing costs	Total costs	Cost of missing production, cost of inspection and cost of parts/material	Seeks the lowest price compared to competitors	l_{26}	Manufacturing costs
		Material costs	Average cost per unit			l_{27}	General costs

Table 7.4 (Continued) Dimensions and maturity criteria for lean

Dimensions	Lean maturity assessment criteria by author					Consolidation of maturity criteria	
	Doolen and Hacker (2005)	Bhasin (2008)	Pakdil and Leonard (2014)	Maasouman and Demirli (2015)	Santos and Tontini (2018)		
			Total costs/ total sales		Lowest total production cost, or the highest production capacity	l_{28}	Cost reduction
dl_7 suppliers	Long-term relationships with suppliers		Establish long-term relationship with our suppliers		Involvement of suppliers in the strategic planning of the company	l_{29}	Long-term relationships with suppliers
	Supplier management		Solve problems jointly with suppliers		Involvement of suppliers in the development of new products	l_{30}	Supplier management
	Supplier evaluation		Consider quality as number one criterion in selecting suppliers			l_{31}	Supplier evaluation

(*Continued*)

Table 7.4 (Continued) Dimensions and maturity criteria for lean

Dimensions	Lean maturity assessment criteria by author					Consolidation of maturity criteria
	Doolen and Hacker (2005)	Bhasin (2008)	Pakdil and Leonard (2014)	Maasouman and Demirli (2015)	Santos and Tontini (2018)	
dl_8 inventory		Stock turnover	Inventory turnover rate	Inventory turnover ratio	Inventory turnover ratc	l_{32} Inventory turnover
		Inventory of raw material, WIP[c] and finished goods	Raw material inventory/total inventory and finished goods inventory/total inventory	Inventory control and inventory level reduction		l_{33} Inventory control
				Heijunka		l_{34} Heijunka

[a] OEE = overall equipment effectiveness.
[b] SPC = statistical process control.
[c] WIP = work-in-progress.

7.4.1 Select evaluation criteria and productive areas of the company

Assessing the adoption level of I4.0 and Lean in companies can be done in different ways. However, an efficient way to promote this type of evaluation is through maturity models properly developed for this context. Maturity models are commonly used to measure the maturity level of an organization or process in relation to a desired state (Schumacher et al., 2016). Maturity models help to define the development state through different levels determined by a starting point (lower degree of development) and an end point (higher degree of development) (Becker et al., 2009). According to Parkes and Davern (2011), maturity models are important indicators of the organizational level and can help to create a roadmap for improvement. In this sense, it can be said that maturity models can be used to identify a starting point that allows organizations to start a development process.

In relation to I4.0, the criteria consolidated in Table 7.2 are used to measure the maturity of the areas under study. Similarly, the criteria shown in Table 7.4 are proposed for measuring maturity in Lean. In this sense, it is proposed to use 20 criteria *'i'* grouped in four dimensions *'di'* for the evaluation of maturity in I4.0, and 34 criteria *'l'* grouped in eight dimensions *'dl'* for assessing the maturity in Lean.

The determination of the evaluated productive areas of the company is based on three requirements: (i) experience with LPs, (ii) operational performance control, and (iii) capacity and feasibility for network data exchange. For criterion (i), it is proposed to use areas that have already evidenced the implementation of LPs for at least 1 year. For criterion (ii), we select areas that have implemented indicators to evaluate operational performance, productivity, quality, and delivery. For criterion (iii), we include productive areas capable of exchanging data with an internal network, or that are at least technically feasible to this condition.

7.4.2 Evaluate the dimensions importance for I4.0 and Lean

The importance of the dimensions related to I4.0 (*di*) and Lean (*dl*) is defined through semi-structured individual interviews conducted with professionals working in the company. According to Martins (2010), semi-structured interviews are a method capable of capturing different points of view of individuals from the same organization. As evidenced in previous works, such as Tortorella and Fogliatto (2014) and Tortorella et al. (2017), some requirements are used to select respondents: (i) a leadership position that allows interviewees to view the organization in various aspects and in a holistic way, (ii) have a minimum experience of 10 years in the implementation of Lean in large companies, and (iii) have thorough knowledge in I4.0.

In order to improve the quality of the answers, before the beginning of each interview, interviewees receive a brief explanation, containing a uniform definition about each dimension used in this study. According to Kothari (2004), this practice can avoid misinterpretations, mitigating possible errors in the answers. Respondents then provide answers to the following question: "What is the importance of the four dimensions *di* and the eight dimensions *dl* for improving operational performance in the productive areas?" For each dimension of I4.0 and Lean, the importance level is assigned following a 5-point Likert scale, where 1 represents 'no importance' and 5 refers to 'maximum importance' for the organization under study (see Appendix A). The different scores obtained for the importance level of each dimension are used to calculate their respective median value. We choose to use the median value of interviewees instead of their average value, since unlike the average, the median is not affected by spurious values or in disagreement with the trend of other values (Tortorella and Fogliatto, 2014). Finally, in order to consolidate the importance level, values assigned by the interviewees are rescaled on a continuous scale of [0, 1] with a high value reinforcing a greater importance.

7.4.3 *Assess the development level of I4.0 and Lean criteria in the productive areas*

To define the development of the criteria of I4.0 and Lean, some employees of the company under study are surveyed. To participate in this survey, each respondent was required to comply with the following requirements: (i) to have a leadership position in one of the selected productive areas, which allows a broader view of area, (ii) have a minimum experience of 5 years within the productive area, (iii) have a minimum experience of 5 years in Lean implementation, and (iv) be familiar with I4.0 technologies. To obtain respondents with different points of view and backgrounds, a senior engineer from each of the productive areas is also invited to answer the questionnaires. The senior engineer is a professional responsible for the technical support and development of new products in the productive areas.

Respondents provide answers and evidence for questions related to the level of compliance with the indicators intended to evaluate operational performance in terms of productivity, quality, and delivery of the area in which they work. In addition, a questionnaire (see Appendix B) consisting of three parts was developed. The first part consists of questions aimed at assessing the demographic profile of respondents (i.e., age, academic background, among others). The second part of the questionnaire assesses the development level of the 20 criteria for I4.0 *ij* ($j = 1, \ldots, 20$) listed in Table 7.2. This assessment is based on a Likert scale ranging from 1 (undeveloped) to 5 (fully developed). The final part of the questionnaire seeks to measure the development level of the 34 criteria for Lean *lk* ($k = 1, \ldots, 34$) (see Table 7.4). Similarly, a 5-point Likert scale is used, where 1 denoted an 'undeveloped' and 5 'fully developed' criteria. Respondents' median values are used and rescaled onto a continuous scale of [0, 1], with a high criterion value indicating a higher development.

To improve the quality of the research instrument, as suggested by Babbie (1999), a pretest was performed before its application. The pretest was conducted with two area supervisors and one senior engineer. After the pretest, the questionnaire was considered appropriate by researchers and could be applied to other respondents. Finally, to ensure that all selected respondents are able to answer questions related to I4.0, some countermeasures are carried out. All respondents receive a prior explanation of each of the four dimensions *di* and 20 criteria *ij*, ensuring a common understanding and avoiding ambiguity in the answers arising from possible misunderstandings. In addition, respondents are informed that there is no right or wrong answer and that responses will be treated anonymously.

7.4.4 *Rank and prioritize areas for a concomitant implementation of I4.0 and Lean*

The ranking of the productive areas is based on the importance level of the dimensions in I4.0 and Lean and the development level of their criteria. A maturity value for I4.0 (mi_n) and Lean (ml_n) for each of the n productive areas is obtained, where ($n = 1, \ldots, h$) as shown by Equations (1) and (2), respectively:

$$mi_n = di_1 \sum_{j=1}^{3} i_j + di_2 \sum_{j=4}^{10} i_j + di_3 \sum_{j=11}^{14} i_j + di_4 \sum_{j=15}^{20} i_j \tag{1}$$

$$\begin{aligned} mme_n &= dme_1 \sum_{k=1}^{2} me_k + dme_2 \sum_{k=3}^{8} me_k + dme_3 \sum_{k=9}^{17} me_k + dme_4 \sum_{k=18}^{21} me_k \\ &\quad + dme_5 \sum_{k=22}^{25} me_k + dme_6 \sum_{k=26}^{28} me_k + dme_7 \sum_{k=29}^{31} me_k + dme_8 \sum_{k=32}^{34} me_k \end{aligned} \tag{2}$$

Subsequently, maturity values for *mi* and *ml* belonging to the same area n are multiplied, determining a final maturity level (mf_n) for each area n ($n = 1, \ldots, h$) as shown in Equation (3). The maturity level mf_n concomitantly contemplates aspects related to I4.0 and Lean in the different areas.

$$mf_n = mi_n \times mme_n \tag{3}$$

Table 7.5 Characteristic of productive areas

Productive areas	No. of employees	Time of Lean implementation (years)
Forming	111	9
Tooling	39	9
Machining	262	10
Heat Treatment	42	9
Assembly	168	10
Shipment	35	10

To better compare values of mf_n, a standardized index (nmf_n) is determined, which is given by the number of standard deviations (Smf_n) of each individual value of mf_n in relation to the average final maturity ($\overline{mf_n}$) as shown by Equation (4) and values higher than ±1.0 are highlighted in Table 7.5. According to Hägg (2003), the standardized score is generally applied in the analysis of maturity level.

$$pmf_n = \frac{mf_n - \overline{mf_n}}{\underline{Smf}_n} \tag{4}$$

A high pmf_n value denotes a higher level of maturity, which means a greater readiness for concomitantly implementing both I4.0 and Lean. To prioritize company's efforts in a more assertive way, productive areas are ranked based on their pmf_n value. In this sense, the pmf_n value should be carefully analyzed, as it can assist the organization in the preparation of plans, definition of priorities, and actions aimed at a concomitant implementation of I4.0 and Lean.

7.5 Results

7.5.1 Company description

The proposed methodology was applied in a large multinational auto parts manufacturer located in Southern Brazil with 59 years of existence and approximately 1,000 employees. The company's productive sector is composed by six large areas, organized into cells and assembly lines. The company acts as a supplier to major automotive manufacturers in the Original Equipment Manufacturer (OEM) and Independent Aftermarket (IAM) segments, supplying products to more than 60 countries.

Seeking to reduce waste, and increase quality and competitiveness, the organization initiated the Lean implementation in 2006. Since then, a large number of initiatives took place, such as workshops, trainings, kaizens, creation of a Lean week, and the establishment of a formal internal Lean training program with a 50-hour workload.

Due to its good results, the company has been recognized by other organizations and specialized entities as a reference in Lean implementation. In terms of I4.0, the organization began its implementation in 2017, and started to gain more knowledge on the subject, conducted training, workshops, and partnered with consultancy in I4.0. In addition, an internal committee was established to discuss issues and prepare studies related to I4.0 adoption.

Finally, the company defined as one of its objectives the implementation of I4.0, which should be approached in synergy with Lean principles and practices in order to maximize the benefits of both approaches.

7.5.2 Selection of productive areas

The productive sector of the organization is structured in six different areas, namely (i) Forming, (ii) Tooling, (iii) Machining, (iv) Heat Treatment, (v) Assembly, and (vi) Shipment. These areas are organized into production cells and assembly lines, have a total of 657 employees, and an average experience of 9.5 years with Lean implementation, as shown in Table 7.5.

Regarding the selection of productive areas, it was based on three requirements. The first requirement consisted in selecting areas with LPs implemented for a minimum of 1 year. Thus, some evidence was gathered to justify that the areas were in accordance with the proposed requirements, such as training records regarding the use of LPs in the mentioned areas, audit reports, and management indicators. In addition, it should be noted that the evaluation of the LPs was carried out with the company's continuous improvement coordinator, who is responsible for implementing the Lean in the company.

All six areas presented evidence of implementation of various practices. The areas with the largest number of LPs implemented were Machining and Assembly with a total of 10 and 8 LPs, respectively. The area with the smallest number of LPs implemented was Tooling (5). All six areas had 43 LPs implemented, which may be justified by the time of Lean implementation in the company (i.e., more than 10 years). It is worth mentioning that '5S', 'kaizen', and 'error-proofing' were present in all areas. 'VSM' (value stream mapping) and 'pull production' were found only in two areas.

With regards to operational performance indicators management, reports were analyzed based on in loco visits. All six areas presented productivity indicators, namely OEE (overall equipment effectiveness) and setup time. Regarding indicators related to quality and delivery, only five areas were able to present evidence of them. Particularly,

Tooling had no indicators related to these two indicators. Thus, this area was suppressed from the selection process that proceeded with the remaining five areas.

Finally, with regards to the capacity and feasibility for network data exchange, the company's maintenance and ICT sectors classified the areas into two categories: (i) area capable of exchanging data or with high feasibility of adequacy and (ii) area not capable of exchanging data and of low feasibility of adequacy. Four aspects were observed to assign into these categories: (a) existence of an available Internet network in the area, (b) equipment with supervisors capable of exchanging data on a network; (c) equipment with actuators controlled by system; and (d) existence of a project that promotes the exchange of area data with some I4.0 technology. Evidence (a), (b), and (c) were observed in all five areas. Evidence (d) was found only in Machining and Shipment. However, the existence of evidence (a), (b), and (c) was considered sufficient to meet this requirement. Thus, all five areas were classified capable of exchanging data or with high feasibility of adequacy. The selected areas as well as the evidence for the requirements are presented in Table 7.6.

7.5.3 *Profile of respondents involved*

To define the importance of the *dl* and *di* dimensions, five managers, responsible for various areas of the organization, were invited to participate in the interviews. Interviews started with an explanation of the dimensions of I4.0 and Lean. Respondents had an average age of 43 years old, an average time of 8 years in this role, and an average experience of 14 years with Lean implementation in large companies. In addition, three of these managers were responsible for all productive areas. The complete demographic profile of respondents is shown in Table 7.7. Interviews lasted approximately 35 minutes and were conducted in April 2019. Observations and comments from managers were written during interviews.

Regarding the evaluation of the importance of the i_{jn} and l_{kn} criteria in the productive areas, questionnaires were applied with 40 respondents. Similar to interviews, respondents initially received an explanation of the criteria of I4.0 and Lean. The sample consisted of 35 leaders (86%) and 5 senior engineers (14%). Leaders had an average experience of 8 years within their respective areas, 9 years with Lean implementation, and led between 28 and 45 employees. Senior engineers had an average experience of 12 years in the productive areas and 11 years with Lean implementation. Because areas had different numbers of employees and leaders, the number of respondents between areas also varied according to the number of leaders. Questionnaires were applied in April 2019, and the

Table 7.6 Requirements for area selection

Requirements	Evidence of productive areas				
	Forming	Machining	Heat Treatment	Assembly	Shipment
LPs implemented in the area for a minimum of 1 year	*Kanban*, Poka-yoke, *kaizen*, *andon*, Heijunka, 5S e TPM	*Kanban*, VSM, Poka-yoke, *kaizen*, Pull production, *andon*, SMED, Heijunka, 5S e TPM	Poka-yoke, *kaizen*, *andon*, Heijunka, 5S e TPM	*Kanban*, Poka-yoke, *kaizen*, Pull production, *andon*, SMED, Heijunka e 5S	*Kanban*, VSM, Poka-yoke, *kaizen*, Pull production, *andon*, 5S e TPM
Performance indicators of productivity, quality, and delivery	OEE of Forming presses, setup time, scrap and rework level, percentage of on-time deliveries	OEE of machining cells, setup time, scrap and rework level, percentage of on-time delivery	OEE of the heat treatment Furnaces, setup time, scrap and rework level, percentage of on-time deliveries	OEE of the assembly lines, setup time, scrap and rework level, percentage of on-time delivery	Individual productivity per employee, scrap and rework level, percentage of on-time deliveries
Capacity to exchange data with an internal network or technical feasibility of adequacy	a. Available Internet network b. Equipment with supervisors capable of exchanging data on a network c. Equipment with actuators controlled via the system	a. Available Internet network b. Equipment with supervisors capable of exchanging data on a network c. Equipment with actuators controlled via the system d. Existence of project that promotes data exchange with some I4.0 technology	a. Available Internet network b. Equipment with supervisors capable of exchanging data on a network c. Equipment with actuators controlled via the system	a. Available Internet network b. Equipment with supervisors capable of exchanging data on a network c. Equipment with actuators controlled via the system	a. Available Internet network b. Equipment with supervisors capable of exchanging data on a network c. Equipment with actuators controlled via the system d. Existence of project that promotes data exchange with some I4.0 technology

OEE = overall equipment effectiveness; VSM = value stream mapping.

Table 7.7 Demographic profile of managers

Demographic profile	Manager 1	Manager 2	Manager 3	Manager 4	Manager 5
Age	47	40	53	40	33
Academic background	MSc	MBA	MBA	MBA	MBA
Areas	Conformation Machining, Heat Treatment, Assembly, and Maintenance	Purchasing and foreign trade	Logistics, PCP, and Shipping	Process Engineering, Tooling, Quality	Product Engineering, Research & Development
Time in the current role (years)	15	11	8	4	2
Lean implementation experience (years)	11	15	21	12	10

Table 7.8 Characteristics of respondents

Area	Number of respondents	Percentage (%)	Characteristics (average)	Years
Forming	8	20.0	Age	38
Machining	12	30.0	Time in the current company	16
Heat Treatment	6	15.0	Time in the current function	9
Assembly	8	20.0	Lean implementation experience	9
Shipment	6	15.0		

average response time was approximately 30 minutes. The characteristics of respondents are presented in Table 7.8.

7.5.4 *Evaluation of the maturity level and ranking of the productive areas*

7.5.4.1 *Evaluation of I4.0 dimensions*

The mi_n and ml_n maturity levels were defined based on information obtained through interviews with managers and questionnaires answered by leaders and experts. Table 7.9 presents the mi_n maturity level of the productive areas, the median value for each dimension (Med. *di*) and criteria in I4.0 (Med. l_{jn}), and their respective rescaled values (*di*) and (i_{jn}). Similarly, Table 7.10 shows the maturity level ml_n, the median values for dimensions (Med. *di*) and criteria in Lean (Med. l_{kn}), and their respective rescaled (*dl*) and (l_{kn}) values.

Regarding the importance of the four *di* dimensions, the dimensions with the greatest importance were di_1 (view of future) and di_3 (operations), both obtaining a median value of 4.0. According to Ganzarain and Errasti (2016), di_1 is dedicated to defining a vision of I4.0 tailored to the organization that, considering its existing strengths and resources, develops its own understanding of I4.0. Understanding the current context, as well as defining an appropriate future vision is critical for the organization to be successful in I4.0 implementation (Geissbauer, 2016; Schuh et al., 2017). di_3 dimension encompasses key concepts of I4.0, such as decentralization, virtualization, analysis, and real-time operations. Once the organization has defined a future vision for I4.0, it becomes possible to define which concepts and technologies are best aligned with it. The proper application of these concepts allows a higher level of communication and interaction among value stream participants (Lasi et al., 2014; Posada et al., 2015).

Regarding the dimensions di_2 (culture) and di_4 (technology and data analysis), both obtained a score of 3.0, and were the least important ones.

Table 7.9 Maturity level *mi*

I4.0 dimensions (*di*)		Med. *di*	*di*	Criteria I4.0 (i_{jn})		Productive areas									
						Forming		Machining		Heat Treatment		Assembly		Shipment	
						Med. i_{j1}	i_{j1}	Med. i_{j2}	i_{j2}	Med. i_{j3}	i_{j3}	Med. i_{j4}	i_{j4}	Med. i_{j5}	i_{j5}
di_1	View of future	4.0	0.80	i_{1n}	New business plan	1.0	0.010	1.5	0.015	1.0	0.010	1.0	0.010	10	0.010
				i_{2n}	Understanding of I4.0 context	1.5	0.015	2.5	0.025	2.0	0.020	2.0	0.020	2.0	0.020
				i_{3n}	Available resources	1.0	0.010	2.0	0.020	2.0	0.020	2.0	0.020	1.5	0.015
di_2	Culture	3.0	0.60	i_{4n}	Knowledge sharing	2.0	0.020	1.0	0.010	1.0	0.010	1.0	0.010	2.0	0.020
				i_{5n}	Collaboration	1.0	0.010	1.5	0.015	2.0	0.020	1.0	0.010	2.5	0.025
				i_{6n}	Flexible culture	1.5	0.015	2.0	0.020	2.0	0.020	2.0	0.020	2.5	0.025
				i_{7n}	I4.0 training	2.0	0.020	2.0	0.020	1.0	0.010	1.0	0.010	1.0	0.010
				i_{8n}	I4.0 leadership	2.0	0.020	2.0	0.020	1.5	0.015	1.0	0.010	1.5	0.015
				i_{9n}	Decision support	1.0	0.010	2.0	0.020	1.0	0.010	1.0	0.010	1.0	0.010
				i_{10n}	I4.0 coordination	1.0	0.010	1.5	0.015	1.0	0.010	1.0	0.010	2.0	0.020

Table 7.9 (Continued) Maturity level *mi*

I4.0 dimensions (di)		Med. di	di	Criteria I4.0 (i_{jn})		Productive areas: Forming Med. i_{j1}	Forming i_{j1}	Machining Med. i_{j2}	Machining i_{j2}	Heat Treatment Med. i_{j3}	Heat Treatment i_{j3}	Assembly Med. i_{j4}	Assembly i_{j4}	Shipment Med. i_{j5}	Shipment i_{j5}
di_3	Operations	4.0	0.80	i_{11n}	Decentralization	1.0	0.010	1.0	0.010	2.0	0.020	1.0	0.010	2.0	0.020
				i_{12n}	Virtualization	1.0	0.010	1.0	0.010	1.0	0.010	1.0	0.010	1.0	0.010
				i_{13n}	Capacity analysis	1.0	0.010	1.5	0.015	1.0	0.010	2.0	0.020	1.0	0.010
				i_{14n}	Real time operations	1.0	0.010	2.0	0.020	2.0	0.020	2.0	0.020	2.0	0.020
di_4	Technology and data analysis	3.0	0.60	i_{15n}	ICT and data security	1.0	0.010	1.0	0.010	2.0	0.020	1.0	0.010	1.0	0.010
				i_{16n}	I4.0 technologies	1.0	0.010	1.0	0.010	1.0	0.010	1.0	0.010	2.0	0.020
				i_{17n}	Advanced machines	1.0	0.010	1.5	0.015	1.0	0.010	1.0	0.010	1.0	0.010
				i_{18n}	Flexibility	1.5	0.015	1.5	0.015	1.0	0.010	2.0	0.020	2.0	0.020
				i_{19n}	Artificial intelligence	1.0	0.010	1.0	0.010	1.0	0.010	1.0	0.010	1.5	0.015
				i_{20n}	Digitalization	1.0	0.010	1.0	0.010	1.0	0.010	1.0	0.010	2.0	0.020
		mi level				0.162		0.206		0.187		0.178		0.216	

Table 7.10 Degree of maturity *ml* by area

Lean dimensions (*dl*)		Med. *dl*	*dl*	Criteria Lean (l_{kn})		Productive areas: Forming Med. me_{k1}	Forming me_{k1}	Machining Med. me_{k2}	Machining me_{k2}	Heat Treatment Med. me_{k3}	Heat Treatment me_{k3}	Assembly Med. me_{k4}	Assembly me_{k4}	Shipment Med. me_{k5}	Shipment me_{k5}
dl_1	View of future	4.0	0.80	mek_{1n}	Strategic planning	3.0	0.018	3.5	0.021	3.5	0.021	3.0	0.018	3.0	0.021
				mek_{2n}	Business plan	3.0	0.018	4.0	0.024	3.5	0.018	3.0	0.021	3.0	0.021
dl_2	Culture	4.0	0.80	mek_{3n}	Programs of involvement and recognition	3.5	0.021	3.0	0.018	3.5	0.021	3.5	0.021	3.0	0.018
				mek_{4n}	Team work	3.5	0.021	3.5	0.021	3.5	0.021	3.5	0.018	4.0	0.024
				mek_{5n}	Leadership performance	3.5	0.021	3.5	0.021	3.0	0.018	3.0	0.018	3.5	0.018
				mek_{6n}	Training programs	3.5	0.021	4.0	0.024	3.5	0.021	3.0	0.018	3.0	0.018
				mek_{7n}	Absenteeism	3.0	0.018	3.0	0.018	3.0	0.018	2.0	0.012	3.5	0.021
				mek_{8n}	Turnover	3.0	0.018	3.0	0.018	3.0	0.018	3.0	0.018	3.5	0.021

Table 7.10 (Continued) Degree of maturity *ml* by area

Lean dimensions (dl)		Med. dl	dl	Criteria Lean (l_{kn})		Productive areas: Forming Med. me_{k1}	Forming me_{k1}	Machining Med. me_{k2}	Machining me_{k2}	Heat Treatment Med. me_{k3}	Heat Treatment me_{k3}	Assembly Med. me_{k4}	Assembly me_{k4}	Shipment Med. me_{k5}	Shipment me_{k5}
dl_3	Operations	4.0	0.80	mek_{9n}	Operational performance	3.0	0.018	3.5	0.021	3.0	0.018	3.0	0.021	3.5	0.021
				mek_{10n}	Standardization of processes	3.0	0.018	3.0	0.018	3.5	0.021	3.5	0.021	3.5	0.024
				mek_{11n}	Setups	3.0	0.018	3.5	0.021	3.0	0.018	3.0	0.018	2.5	0.015
				mek_{12n}	OEE	3.0	0.018	3.0	0.018	3.5	0.021	3.0	0.018	3.0	0.018
				mek_{13n}	TPM	2.5	0.015	3.0	0.018	3.5	0.021	3.0	0.018	2.0	0.012
				mek_{14n}	Preventive maintenance	3.0	0.018	3.0	0.021	3.5	0.021	4.0	0.024	2.5	0.015
				mek_{15n}	Safety	4.5	0.026	4.5	0.026	4.5	0.026	4.0	0.024	4.5	0.026
				mek_{16n}	Pull system	3.0	0.018	3.0	0.018	3.5	0.021	3.0	0.018	4.0	0.024
				mek_{17n}	Continuous improvement	3.0	0.018	3.5	0.021	3.5	0.021	3.0	0.018	4.0	0.024
dl_4	Quality	4.0	0.80	mek_{18n}	Quality management	2.5	0.015	3.0	0.018	3.0	0.018	3.0	0.018	3.0	0.018
				mek_{19n}	Scrap and rework	2.5	0.015	3.0	0.018	3.0	0.018	3.0	0.018	3.0	0.018
				mek_{20n}	Quality costs	3.0	0.018	3.0	0.018	3.0	0.018	3.0	0.018	2.5	0.015
				mek_{21n}	Mistake proofing	3.0	0.018	3.0	0.018	3.0	0.018	3.0	0.018	3.0	0.018

(*Continued*)

Table 7.10 (Continued) Degree of maturity *ml* by area

	Lean dimensions (dl)	Med. dl	dl	Criteria	Lean (l_{kn})	Forming Med. me_{k1}	Forming me_{k1}	Machining Med. me_{k2}	Machining me_{k2}	Heat Treatment Med. me_{k3}	Heat Treatment me_{k3}	Assembly Med. me_{k4}	Assembly me_{k4}	Shipment Med. me_{k5}	Shipment me_{k5}
dl_5	Focus on the customer	4.0	0.80	mek_{22n}	Manages customer relationships	3.5	0.021	3.5	0.021	3.5	0.021	4.0	0.024	4.0	0.024
				mek_{23n}	Customer satisfaction	4.0	0.024	4.0	0.024	3.0	0.021	4.0	0.024	3.5	0.021
				mek_{24n}	Customer oriented	2.5	0.015	4.0	0.024	3.0	0.018	3.0	0.018	3.0	0.018
				mek_{25n}	On time delivery	2.5	0.015	4.0	0.024	4.0	0.024	4.0	0.024	4.5	0.026
dl_6	Costs	4.0	0.80	mek_{26n}	Manufacturing costs	3.5	0.021	3.0	0.018	3.0	0.018	3.0	0.018	4.0	0.021
				mek_{27n}	General costs	3.0	0.018	3.0	0.018	3.5	0.021	3.0	0.018	3.0	0.018
				mek_{28n}	Cost reduction	3.5	0.021	3.5	0.021	3.0	0.018	3.0	0.021	4.0	0.021

Table 7.10 (Continued) Degree of maturity *ml* by area

						Productive areas									
						Forming		Machining		Heat Treatment		Assembly		Shipment	
Lean dimensions (*dl*)		Med. *dl*	*dl*	Criteria Lean (l_{kn})		Med. me_{k1}	me_{k1}	Med. me_{k2}	me_{k2}	Med. me_{k3}	me_{k3}	Med. me_{k4}	me_{k4}	Med. me_{k5}	me_{k5}
dl_7	Suppliers	3.0	0.60	mek_{29n}	Long-term relationships with suppliers	3.0	0.021	4.0	0.024	3.0	0.018	3.0	0.018	4.0	0.021
				mek_{30n}	Supplier management	3.0	0.018	3.0	0.018	2.0	0.012	3.0	0.018	4.0	0.024
				mek_{31n}	Supplier evaluation	3.0	0.018	3.0	0.018	2.0	0.012	2.0	0.012	3.5	0.021
dl_8	Inventory	3.0	0.60	mek_{32n}	Inventory turnover	3.0	0.018	3.0	0.018	3.0	0.018	3.0	0.018	3.5	0.021
				mek_{33n}	Inventory control	3.5	0.021	3.5	0.021	3.5	0.021	3.0	0.021	4.0	0.024
				mek_{34n}	Heijunka	3.0	0.018	3.5	0.021	3.0	0.018	3.0	0.018	3.0	0.018
ml level						0.479		0.518		0.496		0.488		0.515	

di_2 encompasses initiatives aimed at the management and development of professionals aligned with I4.0. I4.0's dynamism and its modern technologies oriented toward digitization and flexibility of production systems (Radziwon et al., 2014) require the creation of an organizational culture that promotes collaboration, flexibility, and knowledge sharing. However, the shortage of qualified professionals are obstacles that concern companies (Geissbauer, 2016), and may become a barrier for the adoption of I4.0 (CNI Brazil, 2016). In this regard, it is important that manufacturing organizations are prepared to make investments and trainings to gain access to I4.0.

Regarding di_4, this dimension addresses technological aspects that enable horizontal and vertical digitization throughout the company (Geissbauer, 2016; Schumacher et al., 2016). Thus, it becomes possible to integrate different sectors of the company, such as engineering, production, and quality. This integration enables improvements in productivity, cost, and quality (CNI Brazil, 2016; Tamás et al., 2016). The utilization of I4.0 technologies and data analysis may lead organizations to new business models and ways of working (Almada-Lobo, 2016; Schuh et al., 2017).

7.5.4.2 *Evaluation of Lean dimensions*

Regarding the evaluation of the importance level of the eight *ml* dimensions, the most important ones were dl_1 (view of future), dl_2 (culture), dl_3 (operations), dl_4 (quality), dl_5 (focus on the customer), and dl_6 (costs), all with a median value of 4.0. dl_1 focuses on the organization of strategies, business plan, as well as the implementation itself (Santos and Tontini, 2018). According to Bhasin (2008), this dimension is fundamental for organizations as it is short-, medium-, and long-term oriented. The ability to visualize the future and anticipate needs ahead of competitors is a critical factor for the success of organizations (Shenhar and Dvir, 1996). dl_2 is dedicated to initiatives aimed at performance appraisal, development, and employee recognition. The efficiency of LPs is strongly linked to the level of training employees receive (Yauch and Steudel 2002; Rothstein 2004; Holman et al., 2005; Longoni et al., 2013; Pakdil and Leonard 2014). In this sense, it is important to consider aspects related to training activities to achieve better efficiency rates (Birdi et al., 2008; Pakdil and Leonard 2014). dl_3 encompasses the set of LPs for a better operational management, such as just in time, TPM, and SMED. According to Arbós (2002), these practices have contributed intensely to improve flexibility and accelerate problem solving in productive areas. A higher maturity level in Lean tends to positively influence operating performance (Chavez et al., 2013; Rasi et al., 2015; Santos and Tontini, 2018). dl_4 represents the organization's ability to manage quality performance, prevent failures, minimize scrap, and control quality costs.

This dimension addresses actions to achieve perfect quality. Errors must be detected and corrected when they are generated (Santos and Tontini, 2018). The Lean concept promotes highly efficient production systems that consume less resources and generate higher quality (Mi Dahlgaard-Park and Pettersen, 2009). dl_5 addresses how the organization determines customer requirements and expectations, such as relationship management, customization, and delivery. According to Singh et al. (2010), all actions and plans in organizations have as their ultimate goal greater customer satisfaction and loyalty. Lean implementation fosters a more competitive company. In this sense, adding value to customers is one of the fundamental principles of Lean (Ohno, 1988; Shah and Ward, 2003; Jones and Womack, 2004; Junior and Filho, 2010; Jasti and Kodali, 2015). dl_6 covers the management of direct and indirect costs related to the company's activities. Lean is a production system whose goal is to reduce costs (Ohno, 1988). According to Karlsson and Åhlström (1996), the dimensions that determine the performance of Lean must be related to specific indicators, such as productivity, quality, delivery time, and cost.

Regarding dimensions dl_7 (suppliers) and dl_8 (inventory), both received a score of 3.0 and were less important based on managers' opinion. dl_7 encompasses initiatives aimed at supplier management and development. The relationship between company and supplier must be strategic for Lean (Jones and Womack, 2004). Reliable suppliers enable inventory reduction (Shah and Ward, 2007). However, variations in time and quantity found in supply chains are a major drag on Lean. In this sense, efforts need to be made to move into that direction. According to Contreras and Mast (2001), long-term relationships and supplier evaluation are important aspects for improving the relationship between organizations and their suppliers. dl_8 covers activities related to inventory management and production leveling. According to Zipkin (2000), effective inventory management requires the evaluation of performance indicators such as turnover and level of raw material, work in process, and finished good items. Thus, Lean helps to reduce costs related to inventory control (Masaaki, 1986). Finally, it should be noted that in general the relatively well-balanced importance of Lean dimensions denotes an integrated implementation arising from a greater familiarity with the theme, especially when compared to I4.0.

7.5.4.3 *Evaluation of final maturity and ranking of areas*

The mi_n and ml_n maturity values of the different production areas were multiplied to define the maturity level mf_n as well as their respective standardized value of pmf_n. Shipment and Machining were the areas that presented the highest maturity levels for both I4.0 and Lean. Particularly, Shipment obtained the highest value of *pmf* (1.17) followed by Machining

Table 7.11 Final maturity degree normalized to I4.0 and Lean

Ranking	Productive areas	*mi*	*Ml*	*mf*	*pmf*
1°	Shipment	0.216	0.515	0.111	**1.17**
2°	Machining	0.206	0.518	0.107	0.83
3°	Heat Treatment	0.187	0.496	0.093	−0.16
4°	Assembly	0.178	0.488	0.087	−0.59
5°	Forming	0.162	0.479	0.078	**−1.25**

(0.83), as shown in Table 7.11. It is worth highlighting the high potential of the Shipment area to reconcile both approaches (I4.0 and Lean). This area has a warehouse management system (WMS) capable of recording all tasks and processes information in real time. A large volume of data can be stored and handled by technologies such as IoT, Big Data, and CPS. As suggested by Pagliosa et al. (2018), IoT, CPS, and Big Data are technologies that show a high degree of synergy with some LPs that are already implemented in the Shipment area (e.g., kanban, VSM, and kaizen). Such LPs are associated with the criteria 'pull systems' (ml_{16}) and 'continuous improvement' (ml_{17}), which obtained the highest score in the Shipment area. In addition, Shipment has a physical space that has already undergone several improvements LPs, such as kaizen and VSM. In this sense, the use of technologies capable of analyzing data and identifying patterns, such as IoT and Big Data, contemplated by the dimension 'technology and data analysis' (*di4*), allows new improvements in the layout of the area. Another positive point identified by the research and highlighted by the area supervisors was the fact that Shipment has the lowest turnover and absenteeism rates, represented by the criteria l_7 and l_8, respectively. This result favors the area in other important criteria that were also evaluated with a high maturity level in I4.0, such as 'knowledge sharing' (i_4), 'collaboration' (i_5), and 'flexible culture' (i_6). In terms of Lean, some criteria were also favored, such as 'teamwork' (l_4) and 'process standardization' (l_{10}). Finally, it is noteworthy that such criteria have also been considered important by previous studies in I4.0 (Geissbauer, 2016; Leyh et al., 2016; Schumacher et al., 2016; Schuh et al., 2017), as well as in Lean (Doolen and Hacker, 2005; Bhasin, 2008; Maasouman and Demirli, 2015; Santos and Tontini, 2018).

Regarding the maturity level of Machining, it presented the highest score in all criteria related to the 'view of future', both in I4.0 (i_1, i_2 and i_3) and in Lean (l_1 and l_2). Understanding the current context, as well as defining a correct future vision is essentially a successful I4.0 implementation (Geissbauer, 2016; Schuh et al., 2017). Similarly, with regards to Lean, an adequate future vision is essential (Bhasin, 2008). An evidence with this respect was that Machining has a project to improve data exchange based on some I4.0 technologies. Currently, the area operates in connection with an Intranet network, allowing the storage and sending of real-time information regarding setups, operational performance, and maintenance to different areas of the company, such as logistics, engineering, and maintenance. This project is in line with company's strategic plan. Another reason was that Machining scored high on other important criteria in both I4.0 (e.g., I4.0 'training' (i_7) and 'I4.0 leadership') and Lean (e.g., 'leadership performance' (l_5) and 'training programs' (l_6)). Regarding the evidence presented by the leadership for these results, it can be highlighted that this area has the largest number of hours invested in training per employee. According to Longoni et al. (2013) and Pakdil and Leonard (2014), success in implementing LPs is also related to the level of training employees have. Regarding the criteria in I4.0, we highlight the participation of all the leadership of the Machining and Forming areas in specialized training in I4.0. The training covered some topics related to I4.0 core technologies, integration of manufacturing through I4.0, and use of advanced robotization. This initiative positively favors the area, as modern I4.0 technologies require qualified professionals. In addition, the scarcity of trained professionals is an obstacle that concerns companies that may become, including a limitation to the adoption of I4.0 (Geissbauer, 2016; CNI Brazil, 2016). In this sense, structuring training programs where employees would be receiving knowledge focused on I4.0 is fundamental. For employees to perform their tasks efficiently within increasingly complex environments, appropriate qualification measures are required at the organizational and technological levels (Gorecky et al., 2017). According to Capgemini Consulting (2015) and Deloitte (2015), manufacturing companies must empower existing employees to make use of new I4.0 technologies.

Regarding the area with the lowest degree of maturity, Forming was the sector that obtained the lowest score in the *pmf* index (–1.25). In total, the area obtained the lowest score in 15 of 20 (75%) criteria evaluated in I4.0 and in 18 of 34 (53%) criteria evaluated in Lean. Unlike Shipment and Machining, Forming was the area that presented the lowest maturity level in all criteria related to the 'view of future' dimension in both I4.0 (i_1, i_2, and i_3) and Lean (l_1 and l_2). In addition, the area had the lowest maturity for four important Lean criteria, i.e., 'quality performance' (l_{18}), 'scrap and rework' (l_{19}), 'meet customer requirement' (l_{24}), and 'on-time deliveries' (l_{25}). It is worth mentioning that the low maturity levels in criteria l_{19} and l_{25} corroborate with the information presented by the Forming supervisors, about the area not being in compliance with the scrap and rework level indicators and percentage of deliveries on time. Regarding the evidence presented by the leadership for not meeting the quality indicator (associated with criterion l_{18}), it is worth mentioning the difficulty in acting on the root cause of quality problems, defects in critical products or components, and mixing of parts. In terms of on-time delivery indicator (associated with criterion l_{24}), the main evidences were delay in delivery of lots and incorrect quantity of parts. However, previous studies showed that aspects related to the criteria l_{18}, l_{19}, l_{24}, and l_{25} are approached by Lean (Shah and Ward, 2003; Womack and Jones, 2004; Liker and Franz, 2011; Jasti and Kodali, 2015). Regarding the I4.0 maturity, a high maturity level was obtained for criteria 'knowledge sharing' (i_4), 'training focused on I4.0' (i_7), and ' leadership I4.0 '(i_8). Forming has the leadership with the most aligned academic and technical education in relation to I4.0, since three of eight (37%) respondents had training in the ICT field. The training of these leaders is considered favorable, since I4.0 is closely supported by the adoption of new information technologies, increasing the need for companies with such professionals (Stock-Homburg, 2013). Finally, it was evident that the leadership of the Forming area participated in the same specialized training in I4.0 offered to Machining.

7.6 Conclusions and future research directions

This work presented a method that allows the assessment of the maturity level for concurrently implementing I4.0 and Lean in the productive areas of manufacturing companies according to their maturity level. The proposed method was illustrated by its application in a large auto parts manufacturer of parts for the automobile industry. Two major contributions can be highlighted in this study. First, in theoretical terms, the proposed method allows to jointly measure the maturity level for I4.0 and Lean in productive areas. This assessment is important due to the fact that many companies, which already have Lean implemented, are starting to adopt I4.0. Nevertheless, organizations generally have difficulty understanding the principles and concepts of I4.0 (Erol et al., 2016; Sanders et al., 2016). Some models in the literature have proposed to assess in isolation the degree of maturity at I4.0 (e.g., Anderl, 2014; Chukwuekwe et al., 2016; Ganzarain and Errasti, 2016; Geissbauer, 2016; Leyh et al., 2016; Schumacher et al., 2016; Tonelli et al., 2016; Leyh et al., 2017; Schuh et al., 2017; Ghobakhloo, 2018). Similarly, maturity models can be found exclusively for Lean (e.g., Doolen and Hacker, 2005; Bhasin, 2008; Hamdi and Leite, 2012; Albliwi et al., 2014; Pakdil and Leonard, 2014; Maasouman and Demirli, 2015; Verrier et al., 2016; Santos and Tontini, 2018). However, the existence of a model that assesses in an integrated and quantitative way the maturity level of both I4.0 and Lean is unknown. Thus, this research complements the body of knowledge on I4.0 and Lean with the proposition of maturity model described.

In practical terms, the integration of I4.0 into management approaches like Lean is still incipient in most organizations. According to Kolberg et al. (2016) and Gjeldum et al. (2016), there is a lack of studies that empirically investigate the relationship between Lean and I4.0. By determining the maturity level in I4.0 and Lean, it is possible to identify the most likely areas of the company to implement both approaches. Thus, the proposed method guides the integration of I4.0 into Lean, providing managers arguments and evidence to prioritize their implementation. Results of this research enable organizations to conduct a more assertive implementation, reducing unnecessary effort and investment.

Regarding the limitations of this study, it is worth highlighting some relevant points. First, the results obtained here cannot be generalizable to the other manufacturing companies, since there are different contexts and specificities in the organizations that need to be observed. It is suggested as an opportunity for future research the application of the proposed method in organizations of different sectors, allowing the verification of its applicability and generalization. Second, the survey assessed the maturity level of both I4.0 and Lean in five productive areas. However, the method did not examine the impact of this association on the operational performance of these areas. In this sense, as much as the maturity level suggests that a given area has a greater readiness to integrate I4.0 and Lean, it cannot be guaranteed that it will also achieve significant improvements in its performance. Future studies may include the operational performance of the areas as a variable to be analyzed, relating it to the identified maturity level.

References

Albliwi, Saja Ahmed, Jiju Antony, and Norin Arshed. "Critical literature review on maturity models for business process excellence." In *2014 IEEE International Conference on Industrial Engineering and Engineering Management*, pp. 79–83. 2014.

Alias, Cyril, Udo Salewski, Viviana Elizabeth Ortiz Ruiz, Frank Eduardo Alarcón Olalla, José do Egypto Neirão Reymão, and Bernd Noche. "Adapting warehouse management systems to the requirements of the evolving era of industry 4.0." In *ASME 2017 12th International Manufacturing Science and Engineering Conference Collocated with the JSME/ASME 2017 6th International Conference on Materials and Processing*, pp. V003T04A051–V003T04A051. American Society of Mechanical Engineers, 2017.

Almada-Lobo, Francisco. "The Industry 4.0 revolution and the future of manufacturing execution systems (MES)." *Journal of Innovation Management* 3, no. 4 (2016): 16–21.

Anderl, Reiner. "Industrie 4.0—advanced engineering of smart products and smart production." In *Proceedings of International Seminar on High Technology*, vol. 19. 2014.

Arbós, Lluís Cuatrecasas. "Design of a rapid response and high efficiency service by lean production principles: methodology and evaluation of variability of performance." *International Journal of Production Economics* 80, no. 2 (2002): 169–183.

Arnold, C., D. Kiel, K. I., and Voigt. "How the industrial internet of things changes business models in different manufacturing industries." *International Journal of Innovation Management* 20, no. 8 (2016):1640015.

Aydos, T. F., and J. C. E Ferreira. "RFID-based system for Lean Manufacturing in the context of Internet of Things." In *IEEE International Conference on Automation Science and Engineering (CASE)*, pp. 1140–1145, 2016.

Babbie, Earl. *Métodos de pesquisas de survey*. Vol. 1. Belo Horizonte: Ed. da UFMG, 1999.

Bagheri, B., S. Yang, H. A. Kao, and J. Lee. "Cyber-physical systems architecture for self-aware machines in industry 4.0 environment." *IFAC Papers Online* 48, no. 3 (2015): 1622–1627.

Bauer, Wilhelm, Bastian Pokorni, and Stefanie Findeisen. "Production assessment 4.0—methods for the development and evaluation of Industry 4.0 use cases." In *International Conference on Applied Human Factors and Ergonomics*, pp. 501–510. Springer, Cham, 2018.

Becker, Jörg, Ralf Knackstedt, and Jens Pöppelbuß. "Developing maturity models for IT management." *Business & Information Systems Engineering* 1, no. 3 (2009): 213–222.

Bhamu, J., and K. Singh Sangwan. "Lean manufacturing: literature review and research issues." *International Journal of Operations & Production Management* 34, no. 7 (2014): 876–940.

Bhasin, Sanjay. "Lean and performance measurement." *Journal of Manufacturing Technology Management* 19, no. 5 (2008): 670–684.

Birdi, Kamal, Chris Clegg, Malcolm Patterson, Andrew Robinson, Chris B. Stride, Toby D. Wall, and Stephen J. Wood. "The impact of human resource and operational management practices on company productivity: a longitudinal study." *Personnel Psychology* 61, no. 3 (2008): 467–501.

Brettel, M., N. Friederichsen, M. Keller, and M. Rosenberg. "How virtualization, decentralization and network building change the manufacturing landscape: an Industry 4.0 perspective." *International Journal of Mechanical, Industrial Science and Engineering* 8, no. 1 (2014): 37–44.

Broy, M., M. V. Cengarle, and E. Geisberger. "Cyber-physical systems: imminent challenges. In *Monterey Workshop*, pp. 1–28. Springer, Berlin, Heidelberg, 2012.

Burch V, F. Reuben, and Brian Smith. "Using simulation to teach lean methodologies and the benefits for Millennials." *Total Quality Management & Business Excellence* 30, no. 3–4 (2019): 320–334.

Capgemini Consulting. Industry 4.0 the Capgemini consulting view. (2015). https://www.de.capgemini-consulting.com/resource-file-access/resource/pdf/capgemini-consulting-industrie-4.0_0.pdf (accessed August 30, 2019).

Chavez, Roberto, Cristina Gimenez, Brian Fynes, Frank Wiengarten, and Wantao Yu. "Internal lean practices and operational performance: the contingency perspective of industry clockspeed." *International Journal of Operations & Production Management* 33, no. 5 (2013): 562–588.

Chen, J. Y., K. C. Tai, and G. C. Chen. "Application of programmable logic controller to build-up an intelligent industry 4.0 platform." *Procedia CIRP* 63, (2017): 150–155.

Chiang, Y., and D. Lee. "Smart manufacturing with the Internet of makers." *Journal of the Chinese Institute of Engineers* 40, no. 7 (2017): 585–592.

Chukwuekwe, Douglas Okafor, Per Schjoelberg, Harald Roedseth, and Alex Stuber. "Reliable, robust and resilient systems: towards development of a predictive maintenance concept within the industry 4.0 environment." In *EFNMS Euro Maintenance Conference*. 2016.

Conger, S., and R. Miller. "Problem-based learning for a lean six sigma course." *Working Papers on Information Systems* 13, no. 1 (2013): 1–9.

Contreras, L., and M. Mast. *Lean Strategies Group LLC*. Bookman, London, 2001.

Dave, B., S. Kubler, K. Främling, and L. Koskela. "Opportunities for enhanced lean construction management using Internet of Things standards." *Automation in Construction* 61 (2016): 86–97.

Deloitte. Industry 4.0 challenges and solutions for the digital transformation and use of exponential technologies. (2015). https://www2.deloitte.com/ch/en/ pages/manufacturing/articles/manufacturing-study-industry-4.html. (accessed August 30, 2014).

Dennis, Pascal. "Produção Lean Simplificada/Pascal Dennis; tradução Rosalia Angelita Neumann Garcia." Bookman, London, 2008.

Dickmann, Philipp, ed. *Schlanker Materialfluss: Mit Lean Production, Kanban und Innovationen*. Springer-Verlag, 2008.

Dombrowski, U., T. Richter, and P. Krenkel. "Interdependencies of Industrie 4.0 & Lean production systems: a use cases analysis." *Procedia Manufacturing* 11 (2017): 1061–1068.

Doolen, Toni L., and Marla E. Hacker. "A review of lean assessment in organizations: an exploratory study of lean practices by electronics manufacturers." *Journal of Manufacturing Systems* 24, no. 1 (2005): 55–67.

Dugenske, A., and A. Louchez. "The factory of the future will be shaped by the Internet of things." *Manufaturing Net* 19, 2014.

Edwards, P., and P. Ramirez. "When should workers embrace or resist new technology?" *New Technology, Work and Employment* 31, no. 2 (2016): 99–113.

Eleftheriadis, M.S., and O. Myklebust. "A guideline of quality steps towards zero defect manufacturing in industry." In *Proceedings of the International Conference of Industrial Engineering and Operations Management*, pp. 332–340. Detroit, USA, 2016.

Erol, Selim, Andreas Jäger, Philipp Hold, Karl Ott, and Wilfried Sihn. "Tangible Industry 4.0: a scenario–based approach to learning for the future of production." *Procedia CiRp* 54 (2016): 13–18.

Ferrera, E., R. Rossini, A. J. Baptista, S. Evans, G. G. Hovest, M. Holgado, E. Lezak, E. J. Lourenço, Z. Masluszczak, A. Schneider, and E. J. Silva. "Toward Industry 4.0: efficient and sustainable manufacturing leveraging MAESTRI total efficiency framework." In *International Conference on Sustainable Design and Manufacturing*, pp. 624–633. Springer, Cham, 2017.

Fleisch, E., M. Weinberger, and F. Wortmann. "Business models and the internet of things." In *Interoperability and Open-Source Solutions for the Internet of Things*, pp. 6–10. Springer, Cham, 2015.

Ganzarain, Jaione, and Nekane Errasti. "Three stage maturity model in SME's toward industry 4.0." *Journal of Industrial Engineering and Management (JIEM)* 9, no. 5 (2016): 1119–1128.

Geissbauer, R., J. Vedso, and S. Schrauf. "Global Industry 4.0 Survey: building the digital enterprise." PwC, 2016.

Ghobakhloo, Morteza. "The future of manufacturing industry: a strategic roadmap toward Industry 4.0." *Journal of Manufacturing Technology Management* 29, no. 6 (2018): 910–936.

Gjeldum, Nikola, Marko Mladineo, and Ivica Veza. "Transfer of model of innovative smart factory to Croatian economy using lean learning factory." *Procedia CIRP* 54 (2016): 158–163.

Gorecky, Dominic, Mohamed Khamis, and Katharina Mura. "Introduction and establishment of virtual training in the factory of the future." *International Journal of Computer Integrated Manufacturing* 30, no. 1 (2017): 182–190.

Gronau, N. "Determinants of an appropriate degree of autonomy in a cyber-physical production system." *Procedia CIRP* 52 (2016): 1–5.

Gubán, Miklós, and György Kovács. "Industry 4.0 conception." *Acta Technica Corviniensis–Bulletin of Engineering* 10, no. 1 (2017): 111.

Hägg, Göran M. "Corporate initiatives in ergonomics—an introduction." *Applied Ergonomics* 34, no. 1 (2003): 3–15.

Hamdi, Olfa, and Fernanda Leite. "BIM and Lean interactions from the bim capability maturity model perspective: a case study." In *IGLC 2012-20th Conference of the International Group for Lean Construction, The International Group for Lean Construction*. 2012.

Hines, P., M. Holweg, and N. Rich. "Learning to evolve." *International Journal of Operations & Production Management* 24, no. 10 (2004): 994–1011.

Holm, M., and P. Ahlstrom. "Lean Service–a literature review." *Center for Innovation and Operations Management* (2010): 1–10.

Holman, David, Stephen Wood, Toby D. Wall, and Ann Howard. "Introduction to the essentials of the new workplace." *The Essentials of the New Workplace* (2005): 1–14.

Illés, B., P. Tamás, P. Dobos, and R. Skapinyecz. "New challenges for quality assurance of manufacturing processes in industry 4.0." *Solid State Phenomena* 261 (2017): 481–486.

Jasti, Naga Vamsi Krishna, and Rambabu Kodali. "Lean production: literature review and trends." *International Journal of Production Research* 53, no. 3 (2015): 867–885.

Jones, Daniel T., and James P. Womack. *A mentalidade enxuta nas empresas: elimine o desperdício e crie riqueza*. Gulf Professional Publishing, 2004.

Junior, Muris Lage, and Moacir Godinho Filho. "Variations of the kanban system: literature review and classification." *International Journal of Production Economics* 125, no. 1 (2010): 13–21.

Kagermann, Henning, Johannes Helbig, Ariane Hellinger, and Wolfgang Wahlster. *Recommendations for implementing the strategic initiative INDUSTRIE 4.0: securing the future of German manufacturing industry; final report of the Industrie 4.0 Working Group*. Forschungsunion, 2013.

Kanigolla, D., E. A. Cudney, S. M. Corns, and V. A. Samaranayake. "Enhancing engineering education using project-based learning for Lean and Six Sigma." *International Journal of Lean Six Sigma* 5, no. 1 (2014): 45–61.

Karlsson, Christer, and Pär Åhlström. "Assessing changes towards lean production." *International Journal of Operations & Production Management* 16, no. 2 (1996): 24–41.

Kern, W., F. Rusitschka, W. Kopytynski, S. Keckl, and T. Bauernhansl. "Alternatives to assembly line production in the automotive industry." In *23rd International Conference on Production Research (ICPR)*, pp. 1–9. Manila, Philipines, 2015.

Kibira, D., K. Morris, and S. Kumaraguru. "Methods and tools for performance assurance of smart manufacturing systems." *National Institute of Standards and Technology, NISTIR*, 8099, 2015.

Kolberg, Dennis, Christoph Berger, B.-C. Pirvu, Marco Franke, and Joachim Michniewicz. "CyProF–Insights from a Framework for Designing Cyber–Physical Systems in Production Environments." *Procedia CIRP* 57 (2016): 32–37.

Kolberg, Dennis, Joshua Knobloch, and Detlef Zühlke. "Towards a lean automation interface for workstations." *International Journal of Production Research* 55, no. 10 (2017): 2845–2856.

Kolberg, Dennis, and Detlef Zühlke. "Lean automation enabled by industry 4.0 technologies." *IFAC–PapersOnLine* 48, no. 3 (2015): 1870–1875.

Kothari, Chakravanti Rajagopalachari. *Research Methodology: Methods and Techniques*. New Age International, 2004.

Kouri, I. A., T. J. Salmimaa, and I. H. Vilpola. "The principles and planning process of an electronic kanban system." In *Novel algorithms and techniques in telecommunications, automation and industrial electronics*, pp. 99–104. Springer, Dordrecht, 2008.

Küsters, D., N. Praß, and Y. S. Gloy. "Textile learning factory 4.0–preparing Germany's textile industry for the digital future." *Procedia Manufacturing* 9 (2017): 214–221.

Landscheidt, Steffen, and Mirka Kans. "Automation practices in Wood product industries: lessons learned, current practices and future perspectives." In *The 7th Swedish Production Symposium SPS, 25–27 October, 2016, Lund, Sweden*. Lund University, 2016.

Lasi, Heiner, Peter Fettke, Hans–Georg Kemper, Thomas Feld, and Michael Hoffmann. "Industry 4.0." *Business & Information Systems Engineering* 6, no. 4 (2014): 239–242.

Lee, Jay, Behrad Bagheri, and Hung–An Kao. "A cyber–physical systems architecture for industry 4.0-based manufacturing systems." *Manufacturing Letters* 3 (2015): 18–23.

Leyh, Christian, Katja Bley, Thomas Schäffer, and Lennart Bay. "The application of the maturity model SIMMI 4.0 in selected enterprises." *Twenty-third Americas Conference on Information Systems*. Boston, 2017.

Leyh, Christian, Thomas Schäffer, Katja Bley, and Sven Forstenhäusler. "Assessing the IT and software landscapes of Industry 4.0-Enterprises: the maturity model SIMMI 4.0." In *Information Technology for Management: New Ideas and Real Solutions*, pp. 103–119. Springer, Cham, 2016.

Liao, Y., F. Deschamps, E. D. Loures, and L. F. Ramos. "Past, present and future of Industry 4.0-a systematic literature review and research agenda proposal." *International Journal of Production Research* 55, no. 12 (2017): 3609–3629.

Liker, Jeffrey K., and James K. Franz. *The Toyota Way to Continuous Improvement: Linking Strategy and Operational Excellence to Achieve Superior Performance*. McGraw Hill Professional, 2011.

Longoni, Annachiara, Mark Pagell, David Johnston, and Anthony Veltri. "When does lean hurt?–an exploration of lean practices and worker health and safety outcomes." *International Journal of Production Research* 51, no. 11 (2013): 3300–3320.

Maasouman, Mohammad Ali, and Kudret Demirli. "Assessment of lean maturity level in manufacturing cells." *IFAC–PapersOnLine* 48, no. 3 (2015): 1876–1881.

Marjani, M., F. Nasaruddin, A. Gani, A. Karim, I. A. Hashem, A. Siddiqa, and I. Yaqoob. "Big IoT data analytics: architecture, opportunities, and open research challenges." *IEEE Access* 5 (2017): 5247–5261.

Marodin, G. A., and T. A. Saurin. "Implementing lean production systems: research areas and opportunities for future studies." *International Journal of Production Research* 51, no. 22 (2013): 6663–6680.

Martinez, F., P. Jirsak, and M. Lorenc. "Industry 4.0. The end lean management." In *10th International days of statistics and economics*, 10, 2016.

Martins, Roberto Antonio. "Abordagens quantitativa e qualitativa." *Metodologia de pesquisa em engenharia de produção e gestão de operações*, pp. 45–61. Elsevier, Rio de Janeiro, 2010.

Masaaki, Imai. *Kaizen: The Key to Japan's Competitive Success*. McGraw-Hill, New York, 1986.

Mi Dahlgaard-Park, Su, and Jostein Pettersen. "Defining lean production: some conceptual and practical issues." *The TQM Journal* 21, no. 2 (2009).

Mora, E., P. Gaiardelli, B. Resta, and D. Powell. "Exploiting lean benefits through smart manufacturing: a comprehensive perspective." In *IFIP International Conference on Advances in Production Management Systems*, pp. 127–134. Springer, Cham, 2017.

Mourtzis, D., A. Vlachou, and V. Zogopoulos. "Cloud-based augmented reality remote maintenance through shop-floor monitoring: a product-service system approach." *Journal of Manufacturing Science and Engineering* 139, no. 6 (2017): 061011.

Mrugalska, B., and M. K. Wyrwicka. "Towards lean production in industry 4.0." *Procedia Engineering* 182 (2017): 466–473.

National Confederation of Industry (CNI) Brazil. "Industry 4.0: a new challenge for Brazilian industry." *CNI Indicators* 17, no. 2 (2016): 1–37.

Netland, Torbjørn, and Kasra Ferdows. "What to expect from corporate lean programs." *MIT Sloan Management Review* 55, no. 4 (2014): 83.

Nightingale, Deborah J., and Joe H. Mize. "Development of a lean enterprise transformation maturity model." *Information Knowledge Systems Management* 3, no. 1 (2002): 15–30.

Ohno, Taiichi. *Toyota Production System: Beyond Large-Scale Production*. CRC Press, Boca Raton, FL, 1988.

Pagliosa, Marcos, Guilherme Tortorella, João Carlos Espíndola Ferreira. "Industry 4.0 technologies and lean practices: a systematic literature survey". *VIII Congresso Brasileiro de Engenharia de Produção*, 2018.

Pakdil, Fatma, and Karen Moustafa Leonard. "Criteria for a lean organisation: development of a lean assessment tool." *International Journal of Production Research* 52, no. 15 (2014): 4587–4607.

Parkes, Alison, and Michael Davern. "A challenging success: a process audit perspective on change." *Business Process Management Journal* 17, no. 6 (2011): 876–897.

Pokorni, Bastian, Sebastian Schlund, Stefanie Findeisen, Arthur Tomm, Dennis Euper, Diana Mehl, Nadine Brehm, Daban Ahmad, Peter Ohlhausen, and Daniel Palm. "Produktionsassessment 4.0." *ZWF Zeitschrift für wirtschaftlichen Fabrikbetrieb* 112, no. 1–2 (2017): 20–24.

Posada, Jorge, Carlos Toro, Iñigo Barandiaran, David Oyarzun, Didier Stricker, Raffaele de Amicis, Eduardo B. Pinto, Peter Eisert, Jürgen Döllner, and Ivan Vallarino. "Visual computing as a key enabling technology for industrie 4.0 and industrial internet." *IEEE Computer Graphics and Applications* 35, no. 2 (2015): 26–40.

Powell, Daryl, Jan Riezebos, and Jan Ola Strandhagen. "Lean production and ERP systems in small- and medium-sized enterprises: ERP support for pull production." *International Journal of Production Research* 51, no. 2 (2013): 395–409.

Radziwon, Agnieszka, Arne Bilberg, Marcel Bogers, and Erik Skov Madsen. "The smart factory: exploring adaptive and flexible manufacturing solutions." *Procedia Engineering* 69 (2014): 1184–1190.

Ranz, F., J. Schuhmacher, and V. Hummel. "Competence development for collaborative work systems in learning factories." In Proceedings of the IIE Annual Conference, pp. 2273–2282. Nashville, USA, 2015.

Rasi, Raja Zuraidah RM, Umol Syamsyul Rakiman, and Md Fauzi Bin Ahmad. "Relationship between lean production and operational performance in the manufacturing industry." In *IOP Conference Series: Materials Science and Engineering*, vol. 83, no. 1, p. 012016. IOP Publishing, 2015.

Ren, L., L. Zhang, F. Tao, C. Zhao, X. Chai, and X. Zhao. "Cloud manufacturing: from concept to practice." *Enterprise Information Systems* 9, no. 2 (2015): 186–209.

Rezende, H. A., C. L. S. Antunes Júnior, P. Senna, L. A. de Souza, and A. Monteiro. "Determination of metrics and risk mapping for the analysis of lean supply chains." *Journal of Lean Systems* 1, no. 1 (2016): 31–50 (in Portuguese).

Rothstein, Jeffrey S. "Creating lean industrial relations: general motors in Silao, Mexico." *Competition & Change* 8, no. 3 (2004): 203–221.

Rüßmann, M., M. Lorenz, P. Gerbert, M. Waldner, J. Justus, P. Engel, and M. Harnisch. "Industry 4.0: The future of productivity and growth in manufacturing industries." *Boston Consulting Group* 9 (2015).

Sackey, S.M., A. Bester, and D. Adams. "Industry 4.0 learning factory didactic design parameters for industrial engineering education in South Africa." *South African Journal of Industrial Engineering* 28, no. 1 (2017): 114–124.

Salmi, A., P. David, E. Blanco, and J. Summers. "Deciding the level of automation during the design of assembly systems: literature review of decision methods and a new approach proposal." In *International Conference on Computers & Industrial Engineering (CIE45)* 2015.

Sanders, Adam, Chola Elangeswaran, and Jens P. Wulfsberg. "Industry 4.0 implies lean manufacturing: research activities in industry 4.0 function as enablers for lean manufacturing." *Journal of Industrial Engineering and Management (JIEM)* 9, no. 3 (2016): 811–833.

Sanders, Adam, Karthik RK Subramanian, Tobias Redlich, and Jens P. Wulfsberg. "Industry 4.0 and lean management–synergy or contradiction?" In *IFIP International Conference on Advances in Production Management Systems*, pp. 341–349. Springer, Cham, 2017.

Santorella, Gary. *Lean Culture for the Construction Industry: Building Responsible and Committed Project Teams.* Productivity Press, New York, 2017.

Santos Bento, Graziela dos, and Gérson Tontini. "Developing an instrument to measure lean manufacturing maturity and its relationship with operational performance." *Total Quality Management & Business Excellence* 29, no. 9–10 (2018): 977–995.

Schmidt, Rainer, Michael Möhring, Ralf-Christian Härting, Christopher Reichstein, Pascal Neumaier, and Philip Jozinović. "Industry 4.0-potentials for creating smart products: empirical research results." In *International Conference on Business Information Systems*, pp. 16–27. Springer, Cham, 2015.

Schuh, Günther, Reiner Anderl, Jürgen Gausemeier, Michael ten Hompel, and Wolfgang Wahlster, eds. *Industrie 4.0 maturity index: Managing the Digital Transformation of Companies*. Herbert Utz, Munich, 2017.

Schumacher, Andreas, Selim Erol, and Wilfried Sihn. "A maturity model for assessing Industry 4.0 readiness and maturity of manufacturing enterprises." *Procedia Cirp* 52 (2016): 161–166.

Seppälä, Pentti, and Soili Klemola. "How do employees perceive their organization and job when companies adopt principles of lean production?" *Human Factors and Ergonomics in Manufacturing & Service Industries* 14, no. 2 (2004): 157–180.

Shah, Rachna, and Peter T. Ward. "Lean manufacturing: context, practice bundles, and performance." *Journal of Operations Management* 21, no. 2 (2003): 129–149.

Shah, Rachna, and Peter T. Ward. "Defining and developing measures of lean production." *Journal of Operations Management* 25, no. 4 (2007): 785–805.

Shenhar, Aaron J., and Dov Dvir. "Toward a typological theory of project management." *Research policy* 25, no. 4 (1996): 607–632.

Simpson, Dayna F., and Damien J. Power. "Use the supply relationship to develop lean and green suppliers." *Supply Chain Management: An International Journal* 10, no. 1 (2005): 60–68.

Singh, Bhim, S. K. Garg, and S. K. Sharma. "Development of index for measuring leanness: study of an Indian auto component industry." *Measuring Business Excellence* 14, no. 2 (2010): 46–53.

Spath, D., S. Gerlach, M. Hämmerle, S. Schlund, and T. Strölin. "Cyber-physical system for self-organised and flexible labour utilisation." In *Proceedings of the 22nd International Conference on Production Research (ICPR)*, vol. 50, p. 22. Iguaçu Falls, Brazil, 2013.

Stock, T., and G. Seliger. "Opportunities of sustainable manufacturing in industry 4.0." *Procedia CIRP* 40 (2016): 536–541.

Strandhagen, J. W., E. Alfnes, J. O. Strandhagen, and L. R. Vallandingham. "The fit of Industry 4.0 applications in manufacturing logistics: a multiple case study." *Advances in Manufacturing* 5, no. 4 (2017): 344–358.

Strueker, J., and H. Weppner. "A cloud-based messaging service for cross-enterprise data exchange with smart objects." *Proceedings of the Eighteenth Americas Conference on Information Systems*, Seattle, Washington, August 9–12, 2012.

Stock-Homburg, Ruth. "Zukunft der Arbeitswelt 2030 als Herausforderung des Personalmanagements." In *Handbuch Strategisches Personalmanagement*, pp. 603–629. Springer Gabler, Wiesbaden, 2013.

Tamás, P., B. Illés, and P. Dobos. "Waste reduction possibilities for manufacturing systems in the industry 4.0." In *IOP Conference Series: Materials Science and Engineering*, vol. 161, no. 1, p. 012074. IOP Publishing, 2016.

Tamás, P. "Decision support simulation method for process improvement of intermittent production systems." *Applied Sciences* 7, no. 9 (2017): 950.

Tao, F., L. Zhang, V. C. Venkatesh, Y. Luo, and Y. Cheng. "Cloud manufacturing: A computing and service-oriented manufacturing model." *Proceedings of the Institution of Mechanical Engineers, Part B: Journal of Engineering Manufacture* 225, no. 10 (2011): 1969–1976.

Tonelli, Flavio, Melissa Demartini, Andrea Loleo, and Chiara Testa. "A novel methodology for manufacturing firms value modeling and mapping to improve operational performance in the industry 4.0 era." *Procedia CIRP* 57 (2016): 122–127.

Tortorella, Guilherme Luz, and Flávio Sanson Fogliatto. "Method for assessing human resources management practices and organisational learning factors in a company under lean manufacturing implementation." *International Journal of Production Research* 52, no. 15 (2014): 4623–4645.

Tortorella, Guilherme Luz, Lizandra Garcia Lupi Vergara, and Evelise Pereira Ferreira. "Lean manufacturing implementation: an assessment method with regards to socio-technical and ergonomics practices adoption." *The International Journal of Advanced Manufacturing Technology* 89, no. 9–12 (2017): 3407–3418.

Tortorella, Guilherme, Ricardo Giglio, and Desirée Hermina van Dun. "Industry 4.0 as a moderator on the relationship between lean and operational performance." In *25th International Annual EurOMA Conference: To Serve, to Produce and to Servitize in the Era of Networks, Big Data and Analytics*. 2018.

Tortorella, G.L., and D. Fettermann. "Implementation of Industry 4.0 and lean production in Brazilian manufacturing companies." *International Journal of Production Research* 56, no. 8 (2018): 2975–2987.

Trstenjak, M., and P. Cosic. "Process planning in Industry 4.0 environment." *Procedia Manufacturing* 11 (2017): 1744–1750.

Verrier, Brunilde, Bertrand Rose, and Emmanuel Caillaud. "Lean and green strategy: the Lean and green house and maturity deployment model." *Journal of Cleaner Production* 116 (2016): 150–156.

Villalba-Diez, J., J. Ordieres-Meré, and G. Nuber. "The HOSHIN KANRI TREE. cross-plant lean shopfloor management." In *Proceedings of the 5th Conference on Learning Factories*, pp. 150–155. Elsevier, Bochum, Germany, 2015.

Wagner, T., C. Herrmann, and S. Thiede. "Industry 4.0 impacts on lean production systems." *Procedia CIRP* 63 (2017): 125–131.

Wan, Hung–da, and F. Frank Chen. "A leanness measure of manufacturing systems for quantifying impacts of lean initiatives." *International Journal of Production Research* 46, no. 23 (2008): 6567–6584.

Wilding, Richard, Beverly Wagner, David M. Gligor, and Mary C. Holcomb. "Understanding the role of logistics capabilities in achieving supply chain agility: a systematic literature review." *Supply Chain Management: An International Journal* 17 no. 4 (2012). doi: 10.1108/13598541211246594.

Womack, James P., and Daniel T. Jones. *A máquina que mudou o mundo*. Gulf Professional Publishing, Houston, TX, 2004.

Xu, Yuchun, and Mu Chen. "Improving just-in-time manufacturing operations by using Internet of Things based solutions." *Procedia CIRP* 56 (2016): 326–331.

Yauch, Charlene A., and Harold J. Steudel. "Cellular manufacturing for small businesses: key cultural factors that impact the conversion process." *Journal of Operations Management* 20, no. 5 (2002): 593–617.

Yoon, J.S., S. J. Shin, and S. H. Suh. "A conceptual framework for the ubiquitous factory." *International Journal of Production Research* 50, no. 8 (2012): 2174–2189.

Zipkin, Paul Herbert. *Foundations of inventory management*. McGraw-Hill, New York, 2000.

Züehlke, D. "SmartFactory – Towards a factory-of-things." *Annual Reviews in Control* 34, no. 1 (2010): 129–138.

Appendix A – Questionnaire for assessing the important level of dimensions in Industry 4.0 and Lean Manufacturing

Applied questionnaire						
This questionnaire aims to know the level of importance of the main dimensions of Industry 4.0 and Lean Manufacturing in your company Thank you very much for your participation!						
(1) Regarding your demographic profile, please indicate						
	Demographic profile					
1	Age					
2	Education background					
3	Role					
4	Sector					
5	Time in current role (years)					
6	Time within company (years)					
7	Experience with Lean implementation (years)					
(2) Regarding Industry 4.0 dimensions, indicate the importance level of each dimension for improving operational performance in the productive areas *Scale (1 = not important, 5 = absolutely important)						
	Dimensions	1	2	3	4	5
1	View of the future					
2	Culture					
3	Operations					
4	Technology and data analysis					
(3) Regarding Lean Manufacturing dimensions, indicate the importance level of each dimension for improving operational performance in the productive areas *Scale (1 = not important, 5 = absolutely important)						
	Lean dimensions	1	2	3	4	5
1	View of the future					
2	Culture					
3	Operations					
4	Quality					
5	Focus on customer					
6	Costs					
7	Suppliers					
8	Inventory					

Appendix B – Questionnaire to evaluate the development level of the criteria for Industry 4.0 and Lean Manufacturing

Applied questionnaire								
This questionnaire aims at identifying the development level of the following criteria for Industry 4.0 and Lean Manufacturing in your company Thank you very much for your collaboration!								
	(1) Regarding your demographic profile, indicate							
	Demographic profile							
1	Age							
2	Educational background							
3	Role							
4	Sector							
5	Time in current role (years)							
6	Time in the company (years)							
7	Experience with Lean implementation (years)							
(2) Regarding Industry 4.0 criteria, indicate their development level in your sector *Scale (1 = underdeveloped, 5 = fully developed)								
Dimensions		Criteria	1	2	3	4	5	
View of the future	1	New business plan						
	2	Understanding of Industry 4.0 context						
	3	Resources availability						
Culture	4	Knowledge sharing						
	5	Collaboration						
	6	Flexible culture						
	7	Training oriented to Industry 4.0						
	8	Industry 4.0 Leadership						
	9	Decision-making support						
	10	Coordination of Industry 4.0						
Operations	11	Decentralization						
	12	Virtualization						
	13	Analysis capacity						
	14	Real-time operations						

Technology and data analysis	15	ICT and data security					
	16	Technologies 4.0					
	17	Advanced machines					
	18	Flexibility					
	19	Artificial Intelligence					
	20	Digitalization					
(3) Regarding Lean Manufacturing criteria, indicate their development level in your sector *Scale (1 = underdeveloped, 5 = fully developed)							
Dimensions		Criteria	1	2	3	4	5
View of the future	1	Strategic planning					
	2	Business plan					
Culture	3	Employees recognition policy					
	4	Teamwork					
	5	Leadership performance					
	6	Training programs					
	7	Absenteeism					
	8	Turnover					
Operations	9	Operational performance					
	10	Process standardization					
	11	*Setups*					
	12	OEE					
	13	TPM					
	14	Preventive maintenance					
	15	Safety					
	16	Pull systems					
	17	Continuous improvement					
Quality	18	Quality performance					
	19	Scrap and rework					
	20	Quality costs					
	21	Failure prevention					
Focus on customers	22	Customer relationship management					
	23	Customer satisfaction					
	24	Meeting customer's requirements					
	25	On-time deliveries					

Costs	26	Production costs					
	27	General costs					
	28	Costs reduction					
Suppliers	29	Suppliers partnership					
	30	Suppliers selection					
	31	Suppliers involvement					
Inventory	32	Inventory turnover					
	33	Inventory control					
	34	Production leveling					

chapter eight

Human resource management in Industry 4.0

A brief trip through literature

Rui Pedro Freitas and Carolina Feliciana Machado
School of Economics and Management, University of Minho, Portugal

Contents

8.1 Introduction

In 2011, at a trade fair in Hannover, Germany, a concept was announced for the first time that seems to be a true business revolution, Industry 4.0 (I4.0), which derives from the German term "Industrie 4.0" and can be labeled in various ways: "Intelligent Factories", "Smart Industry", "Factory of the Future", or "Advanced Manufacturing".

The I4.0 designation connotes the fourth industrial revolution, the successor to three previous industrial revolutions that caused quantum leaps in productivity and changed the lives of people around the world. Klaus Schwab (WEF, 2016), President of the World Economic Forum (WEF), goes further and says that it is the most important, whose ramifications go deeper, than at any other time in the human history.

While many authors predict a sharp job loss, the vast majority of industrial companies assume that work will remain significant in the new era of industrial production, although it will be different. The "Man and Machine in Industry 4.0 – How Will Technology Transform the Industrial Workforce Through 2025?" study, conducted by the Boston Consulting Group on German industry, even estimates that economic growth will stimulate employment growth of around 6% until 2025, with a higher demand for mechanical engineers (10%), generating 390,000 new jobs (Lorenz, Ruessmann, Strack, Lueth, and Bolle, 2015). However, it is highly speculative data and will depend on the form and timing of introduction, of disruptive technologies, into the industry.

The term "I4.0" is applied to a group of rapid transformations in design, production, operation, logistics, and service, producing very significant impacts on all dimensions of the company.

Over the past few years, debate and studies have focus on the promises and benefits that I4.0 would bring to society, both from the market and business perspective.

Consumers benefit in a number of ways, but the three most prominent are perhaps the product's greater customization capabilities, increasingly suited to its preferences and needs; easier access, truly globalized, and digital; and faster, expressed by the factories returning to the consumer market, new technologies employed, and greater control of the entire operational and logistics process.

But there are also benefits to companies, especially those that adapt to the principles of I4.0 and invest in the necessary technologies, responding to market drivers. Businesses should become more responsive, innovative, and efficient, with an opportunity here to improve their performance.

This performance can be measured on the base of various indicators – financial, sales, operational, among others – that depend on both the productivity of equipment and the congruence of business models as well as the workers and organizational models installed. In pursuit of best performance, the relationship between the proliferation of digital technologies and their social consequences is never linear and deterministic. It is a dyad that assumes a reciprocal relationship shaped by the influence of diverse economic, social, and political factors, as is support by Evangelista, Guerrieri, and Meliciani (2014). The so-called *digitization of industry* can be an important step to economic growth and, in what concerns work, to a higher labor productivity, to employment and inclusion, with a strong impact on learning processes, human capital training, and labor demand.

In this context of profound change, many seek to provide insight into the future of the industry by identifying its impacts on the market, work, and organization, which naturally have strong impacts on Human Resource Management Policies and Practices. Thus, this chapter aims to develop a literature review on the theme "Human Resources Management in Industry 4.0":

- It starts by framing I4.0, addressing the concept, technologies, and inherent principles.
- Signals the main expected impacts of I4.0 on the market and industry, work, and organization.
- Addresses recent studies and research on possible human resource policies and practices for I4.0: focusing on I4.0 training and identifying the skills of workers that are most required for I4.0.
- Concludes by giving indications about the timeliness and relevance of further research.

8.2 Industry 4.0

8.2.1 Concepts and evolution

In 2012, answering to a decline in the relative importance of industry in Europe, the European Commission established as target that manufacturing industry should represent 20% of the total EU add value by 2020.

In Europe, 1 in 10 companies is classified as manufacturing industries, which means around 2 million companies, accounting for 33 million jobs. This industry accounts for over 80% of exports and 80% of investments in Research, Development and Innovation (RDI) being one of the key elements of sustainable economic growth. According to Davies (2015), each new job in industry results in the creation of two new jobs in other sectors.

With a digital single market strategy, the European Commission aims to support all industrial sectors in exploring new technologies and managing the transition to a smart and digitized industry, which can boost the productivity and add value of European industries and stimulate growth.

I4.0 thus appears as an attempt of reindustrialization of the Old Europe, focused on improving efficiency, flexibility, and efficient use of resources being characterized by an increasing digitization and interconnection of products, business models, and value chains, many of them disaggregated and in offshoring.[1]

Despite the German origin, the model has spread around the world with the adoption of similar measures in several countries. Tessarini and Saltorato (2016) spell out other global initiatives: the United States, for example, announced in 2011 the *Advanced Manufacturing Partnership* (AMP), which consists of the union between universities, industries, and the federal government to promote investments in rising technologies in the country. In 2014, its successor, *Accelerating US Advanced Manufacturing* (AMP 2.0), seeks to insert a number of additional actions that should be toked to leverage the country's advanced manufacturing capacity. Also noteworthy are programs in China, which introduced *Made in China 2025* in 2015 as a strategy for upgrading the industry in the country, with several targets set for 2020 and 2025. In South Korea, *Korea Advanced Manufacturing System* (KAMS) was created. A project that aims to develop new processes and technologies for the management and integration of production systems.

The European Union (EU) has an outlined strategy for I4.0 that is reflected in its members' policies, with its own agendas.[2]

In Portugal, emanating from the European economic strategies and directives, in recent years, the Government has developed a busy cofinancing agenda for corporate investments focusing on I4.0 core technologies and processes. In the "Portugal 2020" Community Support Framework, business incentive systems value the framework of applications in I4.0, for purposes of merit analysis (typology Productive Innovation), or are – even – a mandatory condition (e-commerce in SMEs Internationalization). Several initiatives are also launched, such as the I4.0 vouchers, with a clear focus in RDi Projects in compliance with I4.0 (Portugal [2018]2020).[3]

[1] Activities of the value chain relocated to other countries.
[2] https://www.innovationpost.it/2017/03/23/digitalizzare-industria-europea/.
[3] https://www.portugal2020.pt/Portal2020.

I4.0 now looks like the holy grail of developed and developing economies. This digitization of the industry will result in a wide range of changes in production, processes, and business models.

In 2015, Hess said that digitization means the process of socioeconomic change triggered by the introduction of technology and digital systems throughout the production network, leading to far-reaching structural changes in business models (quoted by Hirsch-Kreinsen, 2016). The same author distinguishes two stages of digitization. A first phase took place until the late 1990s in sectors of the economy where production, consumption, and communication were the basis of intangible transactions through the use of data and information. A second phase, focusing on digitization of all kinds, which it calls the "second wave", comprises a strong process of changing socioeconomic aspects with the introduction of new technologies.

We are clearly at the crest of this giant wave, where new technologies are bursting.

8.2.2 I4.0 Core technologies

To Baena, Guarin, Mora, Sauza, and Retat (2017), I4.0 depends on a number of innovative technological developments:

- The application of information and communication technology (ICT) to digitize information and integrate systems at all stages of product creation and use (including logistics and sourcing), both within companies and at their borders.
- Cyber-physical systems (CPS) that use ICT to monitor and control physical processes and systems. These may involve intelligent sensors and robots, which can be configured to immediately suit the product being created, or additive manufacturing devices (3D printing).
- Network communications, including wireless and Internet technologies, that serve to interconnect machines with machines, machines with people, systems with systems, and systems with people and products, both within the factory and with suppliers and distributors.
- Simulation, modeling, and virtualization in product design and definition of manufacturing processes.
- Collecting large amounts of data (Big Data) and subsequent analysis and exploration, either immediately on the shop floor or through analysis and cloud computing.
- Increased ICT support for workers, including robots, augmented reality, and smart tools (Figure 8.1).

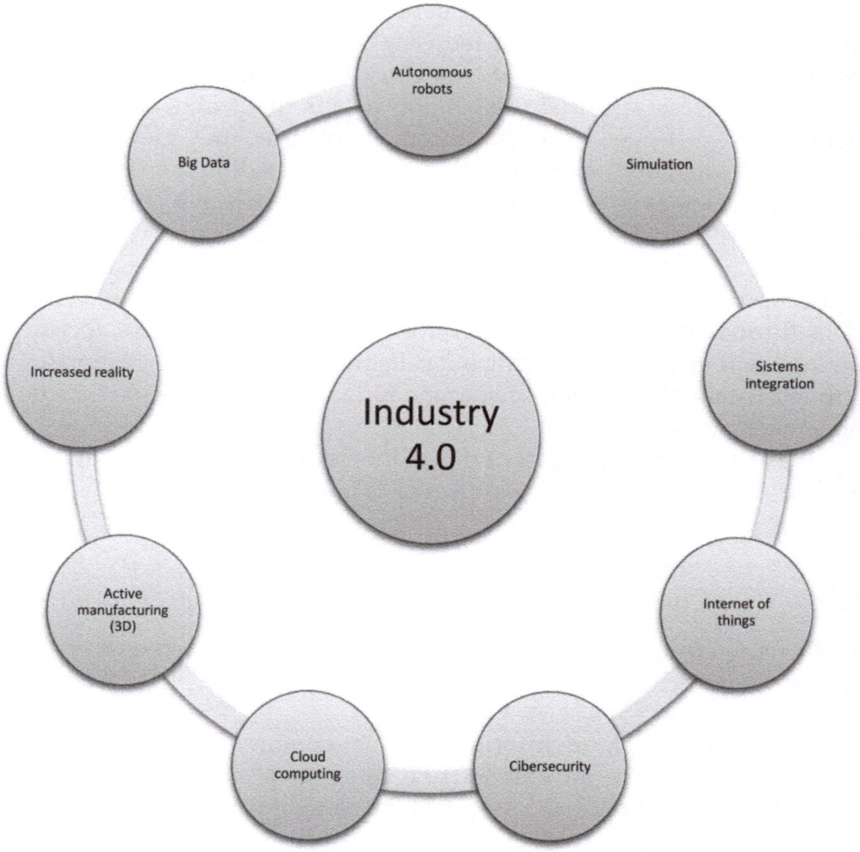

Figure 8.1 The I4.0 nine pillars. (Adapted from The Boston Consulting Group, 2015.)

Briefly, Baena et al. (2017) report that I4.0 makes use of technologies such as the Internet of Things (IoT) and Internet of Services (IoS), CPS, Automation, Information Systems Interaction, Cybersecurity, Big Data, cloud computing, robotics, among others. The diversity of technologies can be aggregated according to their scope of use (Table 8.1)

8.2.3 I4.0 Principles

I4.0 is a collective concept that brings together the organization of technologies and processes in the value chain. Within smart factories, CPS monitors physical processes, creates a virtual copy of the world, and helps to make decentralized decisions. Through IoT, CPS communicate and cooperate with each other and with humans – in real time – and

Table 8.1 Core Technologies from I4.0

Scope of use	Technologies
Production process/ business management	• Digital automation without sensors, use of Programmable Logic Controller (PLC) without sensors • Digital automation with process control sensors • Digital sensor automation with product identification and operating conditions, flexible lines • Collection, processing and analysis of large amounts of business data (Big Data) • Monitoring and remote control of production with systems like MES and SCADA (MES - *Manufacturing Execution Systems; SCADA – Supervisory Control and Data Acquisition*) • Additive manufacturing, collaborative robots (cobots) • Intelligent management systems such as M2M communication (machine-to-machine), Digital Twin and Artificial intelligence (AI)
Product Development	• Integrated engineering systems for product development and manufacturing • Rapid prototyping, 3D printing and similar • Simulations/analysis of virtual models (finite elements, computational fluid dynamics, etc.)
Product / New Business Models	• Collection, processing and analysis of Big Data about the market; monitoring of product use by consumers • Using product-associated cloud services • Incorporating digital services into products (Internet of Things or product service systems)

Source: Author adaptation.

through IoS, services are offered and used by both internal and external value chain participants.

Based on a literature review, Hermann, Pentek, and Otto (2015) identified six principles for I4.0 implementation in the context of a Smart Factory: interoperability, virtualization, decentralization, real-time capability, service orientation, and modularity.

8.2.3.1 Interoperability

Interoperability means that all CPS within the production unit (part supports, assembly unit, and products) are capable of communicating with each other through open networks and common semantic (IoT).

8.2.3.2 *Virtualization*

Virtualization means that CPS is able to monitor physical processes. It presupposes a virtual copy of smart factories, allowing traceability and remote monitoring of all processes through the numerous sensors spread throughout the factory.

8.2.3.3 *Decentralization*

Growing demand for individual products makes it increasingly difficult to control systems in a centralized manner, being necessary to maintain control of the entire system at any time, as Schlick, Stephan, Loskyll, and Lappe (referred by Hermann et al., 2015) state.

Decentralization refers to the possibility that decision-making can be done by the CPS, according to the needs of real-time production, ensuring quality and traceability. The machines will receive commands but can provide information about their work cycle.

8.2.3.4 *Real-time capacity*

For organizational tasks, data must be collected and analyzed in real time. The entire production unit is tracked and analyzed holistically and permanently. Thus, the production unit can react to one machine failure and redirect products to another machine.

8.2.3.5 *Service orientation*

Enterprise services, CPS, and humans are available permanently through the use of service-oriented software architectures coupled with the concept of IoS. To Schlick et al. (2014, referred by Hermann et al., 2015) it can be offered both internally and across company boundaries, which means that the operation of the product-specific process can be composed based on customer-specific requirements in permanent connectivity.

8.2.3.6 *Modularity*

Modular systems are able to adapt in a flexible way to the changing needs by replacing or expanding individual modules. Modular systems can be easily adjusted in case of seasonal fluctuations or changes in product characteristics. Supported by standardized software and hardware interfaces, the new modules are automatically identified and can be used immediately through IoS.

As can be deduced, these principles affect the way the product is manufactured and the organizational context of companies, and can have an impact on customers' perception of the value of the product by increasing a number of benefits valued by the market.

8.3 Impacts of I4.0

8.3.1 Market and industry impacts

According to Bauer, Hämmerle, Schlund, and Vocke (2015), with I4.0, completely new approaches are needed to create products and services that enable comparable standards of prosperity on a global scale, in the future, consuming, radically, and fewer resources. In fact, I4.0 is based on a willingness to reshoring and backshoring some value chain activities around the world, bringing them closer to consumer markets, as focused by Fratocchi et al. (2014). As automation should enable a reduction in labor costs, one of the most important factors in the cost component of many products, relocating production units in developed countries will allow for greater physical proximity to the customer and a reduction in costs and transport environmental impact. In addition, the products will enable electronic identification and monitoring of the entire product life cycle, tracking all processes and chemicals involved in their production as well as their use and disposal, ensuring the opportunity to close the cycle with its recycling contributing to sustainability (VDI and ASME, 2015).

Bauer et al. (2015) state that industry will have to adapt to the growing dynamism of sales markets and the radical challenges that the demand for innovation imposes, notably as regards energy and resource efficiency, and the growing hybridization of products using mechatronics, software, and services.

Intelligent factories allow greater flexibility in production. Production automation, the transmission of data about a product as it passes through the production chain, and the use of configurable robots provide an increased variety of products that can be produced in the same production unit.

According to the same authors, in addition to variety, customer connectivity through IoT and IoS allows for greater customization of products, meeting specific customer needs. This customization will allow the production of small batches and unique products, by the ability to quickly, configure the production equipment.

Customers will also be more involved in the product design process, having the ability to monitor their development and quickly suggest changes. In this respect, as highlighted by Hermann, Pentek, and Otto (2015), virtual modeling and additive production play an important role, providing incentives for customization and innovation, allowing the rapid development of prototypes or new products, which can be produced quickly only with reconfiguration of production lines.

This idea is underlined by Jabbour, Jabbour, Foropon, and Filho (2018) when they point out that the connections between machines, devices, and supply chain elements, with the help of shared information, will create the opportunity for rapid modification, order priorities (as per customer requirements or maintenance requirements), monitor and control assembly line performance, and track deliveries and improve logistics routes.

At I4.0, information becomes a truly additive fuel. Produced data can be collected and analyzed using "Big Data" techniques. Using advanced data analysis, predictive maintenance programs can be defined, avoiding equipment failures and inefficiencies and reducing downtime, with an estimated increase in 20% and 50% of production capacity.

Data-backed supply chains can accelerate production by about 120% in the time required to deliver orders and 70% to get products to market. This reduces the lead time of products, accelerating their deliverability. Another scenario points to the possibility of having fully automated factories with robots, allowing a constant production, remotely manageable, and without loss of quality.

In fact, CPS and new production monitoring technologies can improve product quality by immediately detecting errors or defects and allowing immediate, if not automatic, correction by adjusting production processes in real time. Increasing quality plays an important role in reducing costs and hence increasing competitiveness.

I4.0 will also bring changes in business models. Instead of competing on price, companies can do so on the basis of product and logistical differentiation. IoT and IoS discover new opportunities for companies where coupled service gains enormous preponderance. Shamim, Cang, Yu, and Li (2016) give as example the ability to monitor intelligent products, coupled with performance management and customer interface technological platforms. This trend can help to expand business opportunities and increase revenues.

According to Blunck and Werthmann (2017), this combination of technically sophisticated customer-specific products and short innovation cycles requires companies to exhibit a high degree of dynamism and transformability, only possible through a rapid and efficient cooperation within and between all functions and corporations involved.

The new digital industrial revolution holds the promise of greater production flexibility, mass customization, faster speed, better quality, and better productivity. However, for these benefits to be realized companies will need to invest in equipment, information, and communication technologies (ICT) and data analysis, as well as the integration of data flows throughout the value chain.

The I4.0 study *"The New Industrial Revolution"* (Roland Berger Strategy Consultants, 2015) highlights that this need for integration should occur at the vertical (across all layers of the company) and horizontal levels that extend to all key stakeholders in the industry supply chain. Horizontal scanning will allow smart adaptation to changing circumstances and changing demand and surroundings. Automatically, the optimization of production processes is developed.

8.3.2 Impacts on work and organization

As global competitiveness and the demand for higher performance are increasing, developed economies are investing in technology and process improvement to boost productivity. The relationship of workers with new technologies and processes will lead to profound changes in the organization, the way they work, and their interactions and relationships, which can be aggregated into five axes, as proposed by VDI and ASME (2015) (Figure 8.2).

8.3.2.1 Tasks execution

Following Liau, Deschamps, Loures, and Ramos (2017), the adoption of technological advances should lead to relevant productivity gains,

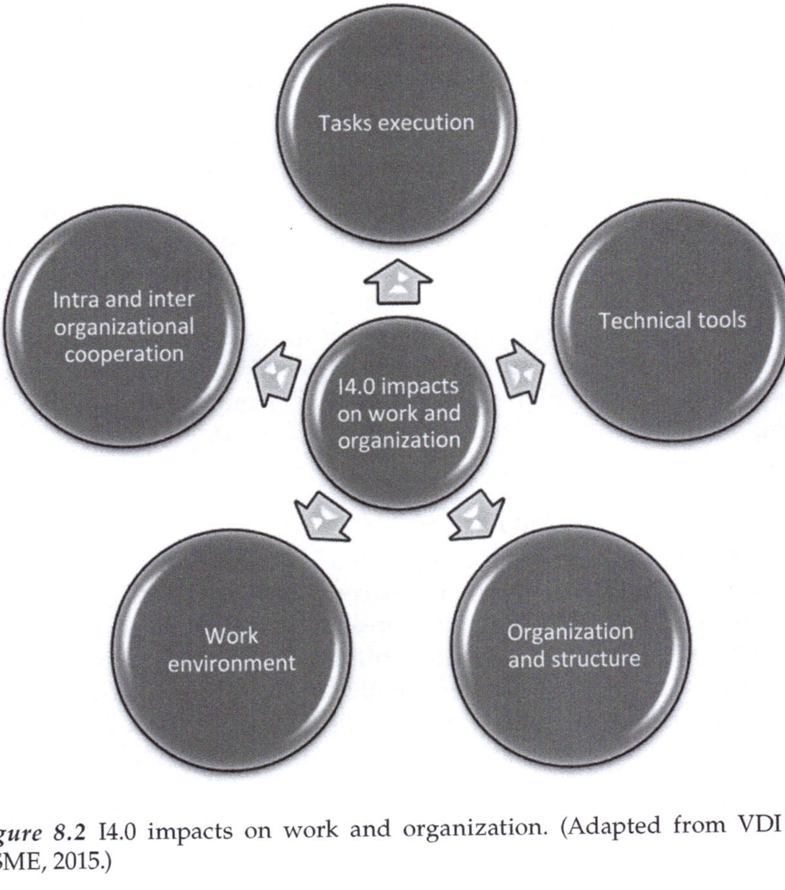

Figure 8.2 I4.0 impacts on work and organization. (Adapted from VDI and ASME, 2015.)

reducing, this way, the number of workers required to maintain or increase productions. Although some work is lost, the level of cooperation between humans and machines will increase significantly. I4.0 will promote significant changes in the way industrial workers perform their tasks and new jobs will be created while others will become obsolete.

Liau et al. (2017) mention that demand will decrease for workers who perform simple and repetitive tasks, because these activities can be standardized and performed by machines, with the introduction of robotics on the shop floor and the computerization of routine work.

However, the experience of the human being is not replaceable. There is a growing awareness, especially in production field, that human capabilities are fundamental and cannot easily be replaced even with industrial automation. Thus, workers will have a greater variety of tasks. As technologies replace a large number of workers' routine tasks, Ramirez (2016) argues that they are free to do more skilled work. Hofmann and Rusch (2017) reinforce this idea by stating that human presence can not only be considered as a type of productive capacity, due to the ability to complete specific work supported by their knowledge and experience, but also as a mechanism for controlling and replacing the system faults. In this case it is based on the ability to face unplanned situations and occasional accidents.

Production processes are changing. They will no longer be organized based on the material flow. On the contrary, they will become organized based on the flow of information, understood as the basic work element. This way, the work involves more process control and supervision tasks, using all kinds of devices and assistance systems. As pointed out by Liau et al. (2017), to perform their job effectively, workers will need to apply a variety of skills, combining know-how related to a specific job or process, with their ability to work with robots or change tools on machines, applying their ICT skills, from the most basic level (using software and interfaces) to the most advanced (advanced programming and analytics).

8.3.2.2 Tools and techniques

The execution of these tasks is supported by a wide range of tools and techniques. Various types of smart devices will be available, such as wearables and tablets, which digitally translate the entire production structure and provide accurate, real-time information, enabling to perform a given task more effectively and efficiently. Intelligent, standalone assistive systems act as interfaces that enable decision-making with fewer errors or risks, improving productivity. Collaborative robotics will be sensitive and intelligent enough to share a workstation with humans, supporting them, requiring, however a high supervisory capacity from them (VDI and ASME, 2015).

In general, the advanced machines, robots, materials, and other devices that form the so-called CPS, through integration with cutting-edge information technologies, develop a new shared productive ecosystem.

According to Chryssolouris, Mavrikios, and Mourtzis (2013), CPS should be understood as highly interactive and sociotechnical systems. A CPS comprises the interaction of information technology between, on the one hand, physical systems with software and, on the other, global data networks with distributed and interactive application systems. Specifically, CPS stands for intelligent and coordinated management of devices, vehicles, processes, and equipment, ranging from production to logistics.

These same authors state that these technological innovations remove the dependence on fixed workplaces and working hours, an idea shared by Rego and Cunha (2018), thus changing the nature of work in both manufacturing and knowledge-intensive occupations, which requires a review of the organizational models.

8.3.2.3 Organization and structure

The adoption of advanced productive technologies leads to what is commonly referred to as the "Factory of the Future", being designed to be highly innovative, networked, and significantly flexible. According to VDI and ASME (2015) this new high-tech production environment will need skilled workers with skills to work with new materials, machines, and especial information. The search for technical talent will drive the shift of job creation within the manufacturing industry by requiring more qualified shop floor staff, trained in multifunctional areas, and able to manage new processes and information systems.

Existing tools and technologies, available to assist skilled work, and the responsibility or job description of factory workers, determine the degree of skilled or manual work, as well as planning and supervision activities.

Henke (2014) points out that this significant change in technologies will represent a significant change in organization and structure. With vertical and horizontal integration, with CPS and IoT, the organization of a factory of the future will be more flexible, changeable, decentralized, and less deterministic compared with today's organizations.

According to Bauernhansl, ten Hompel, and Vogel-Heuser (2014), shop floor teams will be organized so that they can have more autonomy and problem-solving capacity, without much influence from a higher hierarchy (VDI and ASME, 2015). Once the worker has more decision-making power and influence, the organization and structure of companies will become flatter, redefining the bases of authority and decision-making power.

Henke (2014) also states that production work will be allocated on the basis of workers' qualifications and skills, but the new skills that workers experience through their smart devices will increase the possibilities for job rotation and internal mobility, translating greater flexibility.

In a survey conducted by Fraunhofer IPK (Hecklau, Orth, Kidschun, and Kohl, 2017), almost 98% of the 661 participating companies stated that they consider that the flexibility of their production workers will become even more important in the next 5 years. At the same time, more than 72% anticipate that they will have to improve their human resource management (HRM) strategy during that period.

This idea is also highlighted by the WEF (2016), in the Future of Jobs Survey study, in which 44% of respondents indicated that there will be a change in the nature of work, with greater flexibility. For labor-intensive companies, this means that worker flexibility plays an increasingly important role in highly responsive, lean-based[4] supply chains.

In high volatile markets, job rotation, mobility, or transformability will require new tools to strategically manage HR. The term transformability, referred to by Muller, Reinhart, and Spat (2016, quoted by Schuh, Anderl, Gausemeier, ten Hompel, and Wahlster, 2017), derives from discussions about flexibility, which have shown not being the most appropriate term in view of current market requirements. Indeed, flexibility is understood as a quality that is fundamentally intrinsic to a system, while transformability refers to the ability to make rapid and lasting changes to an existing system.

In order that this new system can work with a high performance level, companies need people with decision-making skills, as well as workers with creativity and ICT and technical skills.

In this respect, it is estimated that by 2020 the EU labor market will have a deficit of 825,000 ICT professionals (Davies, 2015). This scarcity has a negative effect on innovation performance, as reported by Tether, Mina, Consoli, and Gagliardi (2005), and may be further accentuated for functions related to advanced manufacturing configurations where large data analysts and cybersecurity experts are needed.

However, despite various initiatives in the EU to encourage eSkills, young people may not necessarily be interested in either industry or workplace digitization (Davies, 2015). On the other hand, demographic developments in developed countries show a decrease in the size of the working-age population by 2025. Adding the psychological detachment of young people from industry highlights the need to implement public compensation programs, some already developed by several countries, such

[4] Lean Thinking is a philosophy that helps the management of an organization. This philosophy is governed by five principles that Womack and Jones (1996) have identified: Value, the Value Chain, the Flow, the Pull system, and Perfection.

as Germany, which shows great openness to welcoming refugees. Thus, it is understood that there is a growing concern among politicians and industry to reverse the policies of replacing older employees by younger employees, scheduling their gradual withdrawal, and implementing mentoring and coaching activities, which will promote cultural change within companies (VDI and ASME, 2015).

The Future of Jobs Survey study (WEF, 2016) also highlights that female talent management is increasingly regarded as a prominent and promising area for workforce planning, as 53% of respondents perceive the promotion of women's participation as a priority item on their organization's leadership agenda. It seems undeniable that, as Rego and Cunha (2018) point out, diversity is critical to improve and contribute to the decision-making process, improve the quality of decisions themselves, and generate creative solutions to address problems and opportunities.

Designing interactions and coordinating diversity will thus be a skill for successful leadership that will have as its ultimate goal to manage the balance of power in the workplace. Hirsch-Kreinsen (2016) adds that it will be perfectly normal to work in mixed teams according to age, gender, origin, and attitudes/approaches, which can potentially trigger major improvements in creativity and productivity, but also represents an increased risk of conflict.

It thus seems clear and inevitable to foresee a reorganization of the labor division that would make it possible to form teams of workers and CPS. In line with the principles of I4.0, opening up a new potential for workers by giving them a greater opportunity to regulate their own work, while requiring more qualifications, namely in the field of media and social skills that impact on their own working environment (Chryssolouris et al., 2013).

8.3.2.4 Work environment

It is very clear that in I4.0, organizations will face many economic, social, and technological challenges that require innovative capabilities and a dynamic workforce. For Shamim et al. (2016), proper management and practices have the potential to improve dynamic capabilities that can lead to innovation, depending on the company's success. If the organization needs to be smart, employees need a good climate for learning and innovation, which requires proper management practices.

All of these challenges involve a continuous process of innovation and learning, which depends on the capabilities of companies and their people, making that management approaches play a vital role in developing dynamic learning capacities that facilitate a climate of innovation. This idea is corroborated by Lorenz et al. (2015), who, also, state that workers will need to be even more open to change, have greater flexibility to adapt to new work environments, and become accustomed to continuous interdisciplinarity and learning. To Rego and Cunha (2018), managing performance means managing the weak-success dichotomy, highlighting them as important pillars of a learning environment.

Thus, the future work environment should be an open and creative space with greater work flexibility. Shift allocations will be more transparent, enabling better planning and autonomy, improving work-life balance, and triggering new work models (VDI and ASME, 2015; Rego and Cunha, 2018).

The workers' own relationship with the company is also changing. On the one hand, by the growing competition of companies for the attraction and retention of talent, which leads to new and aggressive employer branding policies, altering the power relations between workers and companies. On the other, by the way rules, links and interactions are established within the business ecosystem, with an impact on the organizational environment.

According to Bradley, Erickson, Stephenson, and Williams (2000), it is a myth that science and technology will necessarily improve the work experience even because technology allows more freedom, as Ramirez (2016) points out, but also a control and more intense monitoring of workers' performance.

Rego and Cunha (2018) point out this tension between control-freedom stating that companies have to know how to balance these forces, giving freedom within some control, to not prevent the initiative but not create chaos.

IndustriALL Global Union also addresses the huge impact of I4.0 on employment and work, and the risks of new monopolies, dismissals, the spread of precarious digital work, and worker spying, raising the issue of their privacy (Ramirez, 2016). This privacy, as well as the management of the data attached to it, is already framed in the new General Data Protection Regulation (GDPR) that considers, among others, that companies should create a procedure in which the worker (internal or external) is informed about how their personal data will be processed. Namely about the purposes and the period within which they will be stored in order to obtain an informed consent. Still, IndustriALL Global Union emphasizes that I4.0 requires a new social contract, intensifying worker consultation and participation.

8.3.2.5 Intra- and interorganizational cooperation

Cooperation is another feature of I4.0, especially with regard to interactions and relationships.

As Hofmann and Rusch (2017) say, artificial intelligence will allow collaboration and communication between humans and machines, which means a permanent interaction with intelligent technological entities, triggered by voice, touch, or gesture. Teamwork will be central, not only along the vertical (within the organization) but horizontal (along the value chain) levels.

With vertical integration, workers will be given greater responsibility for planning and process improvement. In this regard, Wehinger (2014, quoted by VDI and ASME, 2015) alludes to the fact that networking and interconnection are focal components of I4.0: workers collaborate and communicate without borders both with other workers and machines using smart devices that connect them in real time. Inter organizational communication will develop strongly, which can already be seen in remote digital maintenance, where it is possible to access robotic systems and perform service upgrades, correct errors, or react to a given situation.

At the same time, permanent business-to-business connectivity is increasingly focused on the horizontal integration of information systems through the homogenization of semantics and systems through converged protocols. This enables more co-creation and open innovation, supported by the principles of multidirectional communication between organizations, with virtual access and on-time participation in product development and traceability processes.

With networking supported by IoT and social media, workers gain autonomy and can play new roles in relationships with the various hierarchical levels, or even in contact with other organizations, particularly in environments with a high project focus and knowledge. Project management and networking form effective mechanisms that allow the linking of knowledge, skills, and resources in a more dynamic environment, which raises additional questions about how companies can leverage decision-making and promote greater entrepreneurial freedom. Social Media, in turn, offers the potential to reduce transaction costs in knowledge acquisition, especially those associated with specific projects (Chryssolouris et al., 2013).

According to Rego and Cunha (2018), to some extent opposite, social, and digital interact and must be consistently articulate. Indeed, digital without social results in mechanisms that can embarrass initiative and creativity. On the other hand, social without digital does not allow exploiting the potential of information, decision, and relationship resources underlying the digital.

According to VDI and ASME (2015), digital society extends with the increase of different external parties involved, such as collaborations with research institutes and universities, and classic suppliers of the industry. Cyberspace will become the privileged environment for holding different types of workshops, seminars, and even training sessions, becoming cheaper and more effective due to the knowledge sharing effect and the potential for immediate analysis.

8.4 Human resource management policies and practices for I4.0

The expected impacts at company level and this combination of changes in work and organization will promote new HRM practices, in which new training mechanisms emerge that foresee a conjecturable need for retraining of workers, with the new skills that I4.0 imposes.

This idea is shared by VDI and ASME (2015), which adds that it is necessary to know if companies are preparing themselves to better manage human resources for I4.0. This also means knowing the level of integration of HRM into the strategy of the company, what type of structure will be most common to find or define, and what type of training will be most relevant, as well as develop awareness of the most needed skills.

8.4.1 Human resource management strategies for I4.0

To install the new paradigm of I4.0 in society, understanding the integration of the human being in the productive system is a prerequisite.

According to Lorenz et al. (2015) the system sustainability is highly dependent on the interactions between technology and human beings. Technology should fulfill its function and accommodate workers in this inevitable transformation.

According to the Future of Jobs Survey (WEF, 2016) study, companies will need to build a new approach to workforce planning and talent management. This study shows the main concerns of companies, as shown in Figure 8.3, highlighting the need to invest in the requalification of workers, to support mobility, and job rotation, among other practices.

Figure 8.4 is able to summarize a significant part of the analyzed and reflected literature and is aligned with the five human resource management trends presented by Lorenz et al (2015), which are described below.

8.4.1.1 Adopt new models of work and organization

I4.0 is creating new types of interactions between people and machines that will have significant implications in the nature of work and organizational structure. Enterprises need to rethink decision-making processes and levels of authority, expecting flatter structures and a more decentralized management, promoting innovation, process coordination, and data control rather than trying to automate these critical resources.

8.4.1.2 Develop new approaches to recruitment

Recruitment for I4.0 requires new approaches, more focused on skills and abilities relevant to specific roles, rather than qualifications determined by formal degrees and defined roles. The selection and recruitment process

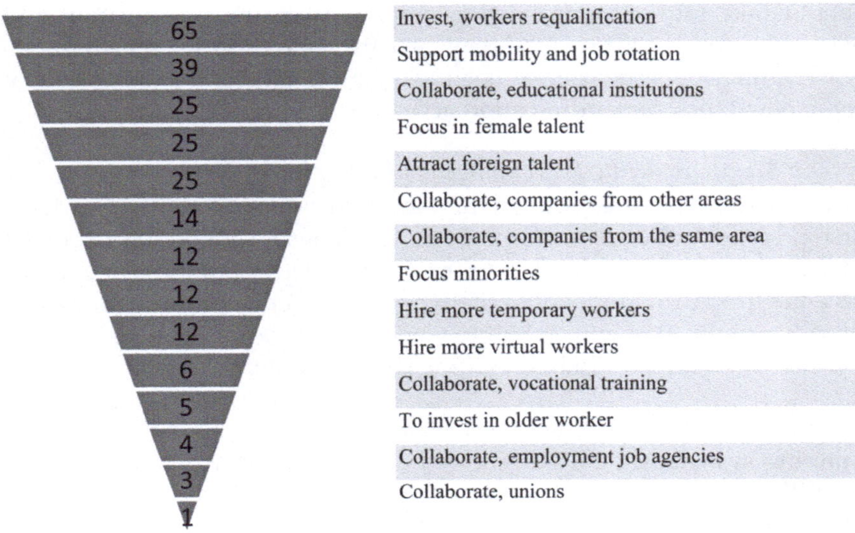

Figure 8.3 HRM strategies. (Adapted from WEF, 2016.)

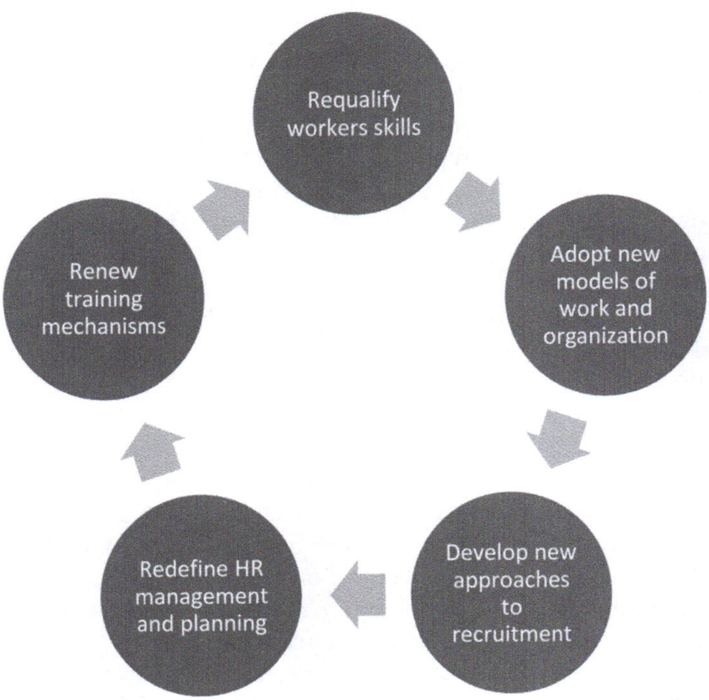

Figure 8.4 Five trends to human resource management. (Adapted from Lorenz et al., 2015.)

should take into account existing employees in the organization and external individuals, provided they have the right skills for specific jobs.

HR managers themselves need to upgrade their skills to implement these new, more demanding approaches.

8.4.1.3 *Redefine HR management and planning*

Strategic workforce planning in this new environment requires new leadership styles and competencies given the intense competition for talent and new forms of work. The desire for productivity improvement and revenue growth requires quantitative (predictive) modeling that can be used internally to gain insight into worker behavior and physical inactivity, and externally to simulate staffing requirements.

8.4.1.4 *Renew training mechanisms*

Training systems should be renewed and aim to provide a broader set of skills and specific work skills, with a particular focus on ICT skills. Hybrid models, which integrate universities and technology centers, are internationally recognized as a superior approach to vocational training and are ideal for building industry-related skills 4.0.

8.4.1.5 *Requalify workers skills*

Companies will need to retrain their employees' skills and knowledge by upgrading their skills to the new I4.0 requirements through effective training programs focused on both job-specific skills and classroom instruction. Training in a broader skill set will often be required given the variety of tasks that will be assigned to workers. A positive perspective on change should be promoted to enable them to adapt to new processes and challenges.

According to Zuboff (1988, referred by Hirsch-Kreinsen, 2016), the improvement of digital qualifications can occur in two ways. On the one hand, the move to a digitized environment is not accessible to all workers. Those who already have higher qualifications and greater usability in virtual contexts will be more prone to their rapid and natural absorption. On the other hand, digitization must be understood as a process of computerization of work that makes possible a greater variety of information about business processes, whose complexity and possible uses can result in new and more complex challenges. To Zuboff (1988, referred by Hirsch-Kreinsen, 2016) skills are based on a theoretical understanding of processes, which lead us to the ability to take in the prerequisites and consequences of using available data at any time.

Rego and Cunha (2018), from their side, add new trends, such as, control, communication, leadership, talent, and learning. Moving on to this learning, it is important to know how to learn, what mechanisms exist, and what changes are taking place in the training of workers.

The reference to new skills and competencies central to I4.0 success is recurring in all types of approach related to task execution, the use of tools and techniques, the organizational structure, the work environment, and intra- and interorganizational relationships.

The next two subsections specifically address the training and skills required for I4.0.

8.4.2 Industry 4.0 training

The Global Risk Report (WEF, 2017), presented at the WEF, highlights that one of the biggest challenges for companies is related with the investment in their human resources training and, according to Chryssolouris et al. (2013), to effectively address the emerging challenges of education and manufacturing and delivery skills, the educational paradigm must be changed.

For this author, people entering the workforce will ideally have all the skills necessary to perform their tasks, but over the life of a working life, the skills and knowledge change rapidly. Continued training of workers to develop new skills and maintain technical skills at a high level becomes an obligation.

As early as 1999, the US Labor Department reported that formal training increased the productivity of production workers by 20% and decreased defect rates by 7% (Mital et al., 1999).

An important point is underlined by VDI and ASME (2015), when it states that qualification measures must address different stages of the educational path from "early education", "school to work transition", and "continuing vocational training". It also highlight that the management of the training needs of each worker must consider a whole life cycle, not necessarily and exclusively focused on the needs of the company but on the symbiosis and balance with the needs of each worker, realizing their lifelong learning.

To Rego and Cunha (2018), lifelong learning cannot be avoided. At this level, these authors highlight three main reasons:

- the rapid technological changes that require new and regular learning,
- longer life span of people who will tend to work more years, and
- the mobility of new generations who have contact with a larger number of organizational contexts.

So, how can continuing learning or training be developed?

Accepting the great diversity of training typologies, Mital et al. (1999) consider that new methods such as initial vocational education and training can be considered, possibly using dual programs. The same authors

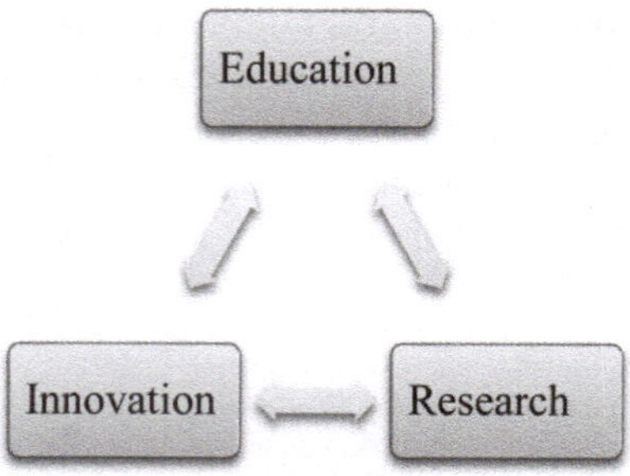

Figure 8.5 The Teaching Factory paradigm. (Adapted from Chryssolouris et al., 2006.)

also highlight internal training, which can be done through online modules, and mentoring, coaching, and seminars, which are a way of ensuring quality information absorption and its practical and incremental application.

However, providing ongoing training is especially challenging for SMEs because the lack of internal resources hinders the ability to create internally generated training structures, even though they can apply to a whole supply of training offer available at technology centers, universities, educational cooperatives, and other industry partners. It will certainly be a challenge to their openness, which will require a new mindset.

As far as I4.0 training is concerned, it is still pertinent to address the concept of "The Teaching Factory". At the conceptual level, Chryssolouris, Mavrikios, and Mourtzis (2006) state that the "Teaching Factory" paradigm (Figure 8.5) is based on a triangle that understands and integrates education, research, and innovation activities, involving industry and academia in a single initiative in order to model the sustainability of companies based on the knowledge produced in universities which, in turn, integrate real-world teaching experience, working methods, and business practice.

Industry contributes to reality, the practical side, while academia contributes to scientific excellence that can be applied to companies. The same authors add that research activities would be intrinsically associated with industrial projects, developing and implementing new technologies and ideas (Figure 8.6).

This symbiosis enables innovation activities to employ knowledge transfer systems to maintain the forefront of the industry, supporting continuous technical improvement through the impregnation of renewed technology, essential for innovation processes.

Within the Teaching Factory, industry operations are communicated to engineering student groups at one or more universities through a web portal, using didactic equipment, with real-time information that considers

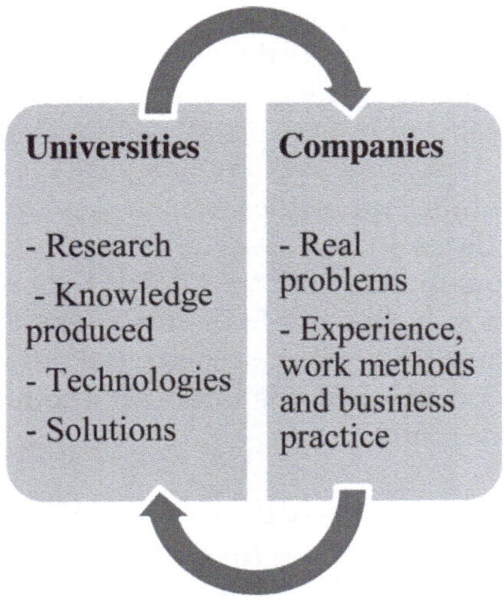

Figure 8.6 Dyad University – Company. (Adapted from Chryssolouris et al., 2006.)

the actual operating systems and facilities of the factory. As an example, we can refer to the Lab-to-Factor project "Systems and Manufacturing and Automation Laboratory", known as KNOW-FACT, launched by a group of organizations, including FESTO, VOLVO, Politecnico di Milano, Technische Universitaet Darmstadt, Tecnalia, and CASP.

Educational activities imbued with the reality of industry help to stimulate curiosity and creativity, support the production of an entrepreneurial spirit, and reduce the psychological gap that is often established between companies and universities.

VDI and ASME (2015), in turn, gives special emphasis to continuous vocational training in the workplace, pointing out that it is important to pay attention to on-the-job training, commonly called, in Portugal, as action-training, referring, however, to other training formats that are considered relevant, namely:

- More specific and results-oriented workshops, capable of effectively assessing both technical and behavioral aspects.
- Professional development courses, as a way of reinforcing technical and personal skills, but with the possibility of being tailored to each worker.
- Industry-university collaboration, in line with The Teaching Factory concept.
- Online courses, sharing problems between departments, contributing to knowledge specialization, and better organizational and process understanding.

- Events such as Open Day and "Presentations" from various departments, which can be developed regularly and allow for greater interdisciplinarity and the strengthening of common commitments to the organization.

The diversity of training typologies, with a view to developing workers' skills and abilities in I4.0, was also addressed in a study by Deloitte (2016) in conjunction with the DMRH Group, which addressed "Corporate Education in Brazil – Skills for a new age of knowledge". Although one should take into account the idiosyncrasies of the Brazilian business framework, it is interesting to highlight five major conclusions (Deloitte, 2016):

1. Companies are adhering more to corporate education practices: The number of companies with teams committed with corporate education increased by 42% in 2016 compared to the survey conducted in 2014. The number of organizations with corporate universities[5] grew by 14%.
2. Training programs get shorter: Although the number of companies with a structure dedicated to corporate education projects has increased, there has been a drop in the number of trained professionals. The overall average hours of training per employee, in turn, fell from 30 (in 2014) to 26 (in 2016).
3. The on-site modality is resumed: On-site training grew from 67% to 74% between 2014 and 2016, while the distance-learning modality fell from 33% to 26% in the same period. The resumption of the face-to-face model suggests that organizations are focusing on traditional practices while preserving investments for the implementation of robust distance education platforms in the future.
4. Partnerships are gaining ground: Corporate education programs developed by in-house or contracted professionals remained at a similar level to the previous survey, with a slight increase in the second modality and a slight decrease in the first.
5. Hard skills grow in comparison to soft skills: Skills considered technical and specific for the development of certain functions (known as hard skills) have gained relevance in relation to skills related to personality aspects (called soft skills). This movement suggests the organizations pragmatism regarding the training needed to increase productivity and efficiency. However, Deloitte (2016) still defends the principle according to which the organizations'

[5] Corporate universities correspond to an education system linked to companies whose main aim consists in developing and educating employees, customers, and suppliers in order to optimize organizational strategies.

investments in soft skills are critical to the business and challenges of today's job market.

This last conclusion puts the emphasis on the type of skills and abilities required for I4.0, which should be seen in more detail.

8.4.3 Skills and competences required in I4.0

According to *The Future of Jobs Employment* study (WEF, 2016), the rapid changes that we are facing in the technological, demographic, and socio-economic fields are going to change not only the organizations and businesses models, but also the abilities and competences required by the employers, shortening the validity and applicability of workers' current skills. Indeed, and following this study, by 2020, on average, more than one third of the core skills needed to perform most functions will be composed of skills not yet considered crucial today.

The adoption of new technologies leads to the need for new skills and competences to perform new tasks, where machines and information systems form a collaborative work ecosystem that redefines the role of workers.

The same study considers that these practical skills will also be subject to accelerating change and significant disruption in the immediate future. On the technical side, new requirements for equipment repair and maintenance, equipment control and operation, programming, data analysis, and quality control stand out. However, the emphasis has mostly been on behavioral skills.

Aires, Moreira, and Freire (2017) analyze several studies, dated from 2015 to 2017, and summarize the skills/competences required for I4.0 (Table 8.2).

8.5 Conclusion

I4.0 is increasingly becoming a reality and an inevitability for companies. Otherwise they will lose their competitiveness and eventually move to extinction. I4.0 is a current phenomenon with tremendous force that poses huge challenges to management and people.

It seems that everything will change, from product and service development, with greater customer participation, and the desired customization; production using an ecosystem made up of machines, robots, information, and people completely connected, both inside and outside the boundaries of the factories of the future; a digitally and informationally integrated value chain; even product delivery is changing, with new flying couriers knocking on doors. The world changes and business models change.

Table 8.2 Competences required for Industry 4.0

Authors	Year	Title/Source	Competences and skills identified and worked
Chen and Zhang	2015	Study on training system and continuous improving mechanism for mechanical engineering. The Open Mechanical Engineering Journal, 9, 7–14.	• Ability to put knowledge into practice and ability to innovate. • Knowledge: mathematics and natural sciences; basic knowledge in mechanics mechanical design, manufacturing machines, and automation. • Skills: independence capacity, analyze and solve engineering problems, communication capacity. • Qualities: healthy personality and physical, social and moral responsibility, spirit of innovation and learning ability
Sorko and Irsa	2016	Engineering education – status quo in Austria in comparison with the academic field of business education. Turkish Online Journal of Educational Technology, 890-894.	• Reproduce simple knowledge. • Analyze, evaluate, and develop situations with creativity and innovation.
CNI	2016	Confederação Nacional da Indústria (CNI). Desafios para a indústria 4.0 no Brasil. Distrito Federal: Brasília.	• Work in multidisciplinary teams. • Have a high level of technical knowledge. • Ability to interact with other areas of knowledge.
Deloitte	2016	Educação Corporativa no Brasil: habilidades para uma nova era do conhecimento. Deloitte.	• This study does not list competences, but refers to the development of technical skills (hard skills), and behavioral skills (soft skills).

Table 8.2 (Continued) Competences required for Industry 4.0

Authors	Year	Title/Source	Competences and skills identified and worked
World Economic Forum	2016	The future of jobs employment, skills and workforce strategy for the fourth industrial revolution. WEF.	Skills: • Cognitive skills: cognitive flexibility, creativity, logical thinking, problem sensitivity, mathematical thinking, and visualization. • Physical skills: physical strength and manual and precision dexterity. Basic competences: • Content competences: active learning, speaking, reading comprehension, written expression, and ICT literacy. • Process competences: active listening, critical thinking, others, and self-monitoring. Transversal competences: • Social competences: team coordination, emotional intelligence, negotiation, persuasion, service orientation, and people training. • Systemic competences: judgment and decision-making and systemic analysis. • Competence to solve complex problems: complex problem solving. • Resource management skills: financial resource management, material resource management, people management, and time management. • Technical competences: repair and maintenance of equipment, control and operation of equipment, and programming and quality control.

(Continued)

***Table 8.2 (Continued)* Competences required for Industry 4.0**

Authors	Year	Title/Source	Competences and skills identified and worked
Garbie	2017	Incorporating sustainability/ sustainable development concepts in teaching industrial systems design courses. Procedia Manufacturing, 8, 417–423.	• Sustainability and sustainable development capabilities.
Veronina and Moroz	2017	Substantiation of foresight research of development strategy of descriptive geometry, engineering geometry and computer graphics departments on the basis of industrial 4.0 ideology. Man in India, 97(3), 375–389.	• Technical knowledge. • Creativity. • Communication.

Source: Adapted from Aires, Moreira, and Freire (2017).

The work also changes, as well as the type of tasks and how they will be performed by machines or humans, or collaboratively by the two. If to the firsts codes and commands are introduced to perform well, to the seconds, to us, the level of demand is immeasurably challenging.

Businesses are required to develop a new plan to retrain workers' skills through new training mechanisms, with a combination of training models focused on people's active development, such as mentoring and coaching, e-learning, workshops, among others, essential for lifelong learning, where the universities' role is also becoming more important as a lever for innovation. New and more flexible, decentralized, and changing models of work and organization will need to be adopted. The way in which they manage their people will be increasingly supported by predictive information and models, supporting mobility and turnover between tasks, but not without the necessary control that accommodate a greater diversity of their workers' profiles and boost the workers' entrepreneurial and creative spirit. Business success will also be driven by their ability to attract and retain talent, thus implying new approaches to recruitment and new leadership styles, more open and democratic.

First and foremost, workers are required to master technological languages that allow them to live with CPS and robots, mastering tools and techniques that enable real-time action, with greater efficiency and effectiveness. People will work in a more collaborative ecosystem, with other people, with machines and information systems that are vertically and horizontally integrated, and new interactions within and with other organizations will emerge. Workers will have greater autonomy and decision-making skills, making cognitive and cross-cutting skills likely to become more important so that they can perform new, less routine but more interdisciplinary tasks that are central to their productivity and business model sustainability, increasingly volatile. Their ability to adapt to new organizational environments is undermined, asking them to be more open to mobility and change. And change is never easy.

There is thus a new way of working, interacting, and relating, which is foreseen in the short term, questioning our knowledge, skills, and abilities.

In a new industrial age, where factory walls are transparent bits immersed in a digitally interconnected and interdependent world, the role of the human being is rethought. The role of the worker has to be reshaped, and the company where he lives has to be rebuilt.

Numbers speak for themselves. Or rather, as the Robotics Online Marketing Team (2018) announces, while we are still understanding the aim of I4.0, "industry 5.0" seems to be in the horizon.

Truly speaking, maybe it's not just new "sound bites". A search of the concept "industry 5.0" on google on January 12, 2019, returned 21,000 results. And less than 2 months earlier, on November 27, 2018, the European Economic and Social Committee (EESC, 2018) launched the theme of the impact of I4.0 on people, which it called "Industry 5.0", embedded in the interaction between creativity and skills of humans and the speed, productivity, and consistency of robots. This vision aims to focus more on the cooperation between robotics and humans, through the combination of their divergent forces, in order to create a more inclusive future, focused on the human being.

At a rapid pace, therefore, there seems to be growing concern about how technology behind I4.0 can help humans and robotics work better. This entails learning more about human-machine collaborative interactions and relationships, with key implications for tasks, income, and work organization. For EESC (2018), the debate has yet to deepen by seeking to answer questions such as what kind of skills we need to develop, taking into account that we work, in a growing way, collaboratively with industrial robots; and how can we benefit, in the best possible way, from intelligence, creativity, and dexterity.

In full agreement with the pertinence of the issues raised, it might be useful to start by developing a research that could contribute to more effectively know the position of Portuguese companies in key issues of I4.0, with particular focus on People Management.

References

Aires, R.W.A.; Moreira, F.K. and Freire, P.A. (2017). Indústria 4.0: Competências requeridas na quarta revolução industrial. International Congress of Knowledge and Innovation – Ciki, [S.l.], 1 (1). Available at http://proceeding.ciki.ufsc.br/index.php/ciki/article/view/314. Accessed on 30 December, 2018.

Baena, F.; Guarin, A.; Mora, J.; Sauza, J. and Retat, S. (2017). Procedia manufacturing, 9. Learning factory: The path to Industry 4.0. 7th Conference on Learning Factories, CLF, Darmstadt, Germany. doi: 10.1016/j.promfg.2017.04.022.

Bauer, W.; Hämmerle, M.; Schlund, S. and Vocke, C. (2015). Transforming to a hyper-connected society and economy – towards an "Industry 4.0". 6th International Conference on Applied Human Factors and Ergonomics and the Affiliated Conferences, AHFE 2015, Las Vegas, United States.

Bauernhansl, T.; ten Hompel, M. and Vogel-Heuser, B. (2014). *Industrie 4.0 in produktion, automatisierung und logistik – anwendung, technologien, migration.* Deutschland: Springer Vieweg.

Blunck, E. and Werthmann, H. (2017). Industry 4.0 – maximizing economic benefits and firm competitiveness for a circular economy. XVI International Business & Economy Conference (IBEC) – Chile 2017. Available at SSRN https://ssrn.com/abstract=2921031. Accessed on 2 October 2019.

Bradley, H.; Erickson, M.; Stephenson, C. and Williams, S. (2000), *Myths at work.* Cambridge: Polity.

Chryssolouris, G.; Mavrikios, D. and Mourtzis, D. (2013). Manufacturing systems: skills & competencies for the future. Forty Sixth CIRP Conference on Manufacturing Systems, Patras, Greece.

Chryssolouris, G.; Mavrikios, D.; Papakostas, N. and Mourtzis, D. (2006). Education in manufacturing technology & science: a view on future challenges & goals. Inaugural Keynote, Proceedings of the International Conference on Manufacturing Science and Technology, Melaka, Malaysia.

Davies, Ron. (2015). *Industry 4.0 – Digitalisation for productivity and growth.* Brussels, Belgium: EPRS | European Parliamentary Research Service.

Deloitte. (2016). Educação corporativa no brasil – habilidades para uma nova era do conhecimento. Available at https://www2.deloitte.com/br/pt/footerlinks/pressreleasespage/pesquisa-educacao-2016.html. Accessed on 30 December 2018.

Evangelista, R.; Guerrieri, P. and Meliciani, V. (2014). The economic impact of digital technologies in Europe. *Economics of Innovation and New Technology* 23(8), 802–824.

EESC – European Economic and Social Committee. (2018). Industry 5.0. Available at https://www.eesc.europa.eu/en/agenda/our-events/events/industry-50. Accessed on 5 January 2019.

Fratocchi, L.; Iapadre, P.; Ancarani, A.; Di Mauro, C.; Zanoni, A. and Barbieri, P. (2014). Manufacturing reshoring: threat and opportunity for east central Europe and Baltic countries. January, 83–118. https://doi.org/10.4018/978-1-4666-6054-0.ch004.

Hirsch-Kreinsen, H. (2016). Digitization of industrial work: development paths and prospects. *Journal of Labour Market Research* 49, 1–14. https://doi.org/10.1007/s12651-016-0200-6.

Hecklau, F.; Orth, R.; Kidschun, F. and Kohl, H. (2017). Human resources management: meta-study – analysis of future competences in Industry 4.0. *Fraunhofer IPK, division corporate management*, Berlin, Germany.

Henke, M. (2014). Management der Industry 4.0 – Heraus-forderungen und Handlungsfelder für die erfolgsorientierte Steuerung von Unternehmen. *Zukunftskon-gress Logistik* 32, 5. Dortmunder Gespräche, Germany.

Hermann, M.; Pentek, T. and Otto, B. (2015). Design principles for industrie 4.0 scenarios: a literature review. Working paper no. 01. https://doi.org/10.13140/RG.2.2.29269.22248.

Hofmann, E. and Rusch, M. (2017). Industry 4.0 and the current status as well as future prospects on logistics. *Computers in Industry* 89, 23–34. https://doi.org/10.1016/j.compind.2017.04.002. Availabe at https://www.innovationpost.it/2017/03/23/digitalizzare-industria-europea/. Accessed on 30 December 2018.

Jabbour A.; Jabbour C.; Foropon, C. and Filho M. (2018). When titans meet – can Industry 4.0 revolutionise the environmentally-sustainable manufacturing wave? The role of critical success factors. *Montpellier Business School*. https://doi.org/10.1016/j.techfore.2018.01.017.

Liau, Y. X.; Deschamps, F.; Loures, E. D. R. and Ramos, L. F. P. (2017). Past, present and future of Industry 4.0—a systematic literature review and research agenda proposal. *International Journal of Production Research* 55(12), 3609–3629. https://doi.org/10.1080/00207543.2017.1308576.

Lorenz, M.; Ruessmann, M.; Strack, R.; Lueth, K. and Bolle, M. (2015). *Man and machine in Industry 4.0: How will technology transform the industrial workforce through 2025?* Boston: Boston Consulting Group.

Mital, A.; Pennathur, A.; Huston, R.L.; Thompson, D.; Pittman, M.; Markle, G.; Kaber, D.B.; Crumpton, L.; Bishu, R.R.; Rajurkar, K.P.; Rajan, V.; Fernandez, J.E.; McMulkin, M.; Deivanayagam, S.; Ray, P.S. and Sule, D. (1999). The need for worker training in advanced manufacturing technology (AMT) environments: a white paper. *International Journal of Industrial Ergonomics* 24(2), 173–184. Available at http://utminers.utep.edu/apennathur/papers/training/whitepaper.pdf. Accessed on 3 January 2019.

Portugal. (2018) 2020. A inovar consigo. Available at https://www.portugal2020.pt/Portal2020. Accessed on 30 December 2018.

Ramirez, P. (2016). When should workers embrace or resist new technology? *New Technology, Work and Employment* 31(2), 99–113. https://doi.org/10.1111/ntwe.12067.

Rego, A. and Cunha, M.P. (2018). Gestão de Pessoas 4.0 – 10 Tendências e desafios [ebook]. Nova School of Business & Economics. Available at http://blog.exed.novasbe.pt/. Accessed on 30 December 2018.

Robotics Online Marketing Team. (2018). What is Industry 5.0 and how will industrial robots play a role? Available at https://www.robotics.org/blog-article.cfm/What-is-Industry-5-0-and-How-Will-Industrial-Robots-Play-a-Role/99. Accessed on 3 January 2019.

Roland Berger Strategy Consultants. (2015). Industry 4.0 – the new industrial revolution. How Europe will succeed. Available at https://www.roland-berger.com/en/Publications/Industry-4.0-%E2%80%93-the-new-industrial-revolution.html. Accessed on 12 November 2018.

Schuh, G.; Anderl, R.; Gausemeier J.; ten Hompel, M. and Wahlster, W. (Eds.). (2017). Industrie 4.0 maturity index. Managing the digital transformation of companies. *Acatech study.The acatech study series*. Munich, Germany: National Academy of Science and Engineering.

Shamim, S.; Cang, S.; Yu, H. and Li, Y. (2016) *Management approaches for Industry 4.0: A human resource management perspective*. IEEE Congress on Evolutionary Computation (CEC), 5309–5316. https://doi.org/10.1109/CEC.2016.774836.

Tessarini, G. J. and Saltorato, P. (2016). Impactos da Indústria 4.0 na organização do trabalho: uma revisão sistemática da literatura. *Revista Produção Online* 18(2), 743–769. Florianópolis, Brasil. http://dx.doi.org/10.14488/1676-1901.v18i2.2967.

Tether, B.; Mina, A.; Consoli, D. and Gagliardi, D. (2005). A literature review on skills and innovation. How does successful innovation impact on the demand for skills and how do skills drive innovation? A CRIC Report for the Department of Trade and Industry. ESRC Centre for Research on Innovation and Competition University of Manchester, UK.

The Boston Consulting Group. (2015). Industry 4.0: The future of productivity and growth in manufacturing industries. Available at https://www.zvw.de/media.media.72e472fb-1698-4a15-8858-344351c8902f.original.pdf. Accessed on 2 October 2019.

VDI (The Association of German Engineers) and ASME (American Society of Mechanical Engineers). (2015). *A discussion of qualifications and skills in the factory of the future: A German and American perspective*. White paper, Düsseldorf, Germany.

WEF (World Economic Forum). (2016). *The future of jobs employment, skills and workforce strategy for the fourth industrial revolution*. Geneva, Switzerland.

WEF (World Economic Forum). (2017). *Global risk report*, 12th ed. Geneva, Switzerland.

Womack, J. P. and Jones, D. T. (1996). *Beyond Toyota: How to root out waste and pursue perfection*. Boston, MA: Harvard Business Review, September-October, 140–158.

Index

Q

R

S

T

U

V

W